The
Inaccessible Earth

The
Inaccessible Earth

G. C. Brown
Open University

A. E. Mussett
Liverpool University

London
GEORGE ALLEN & UNWIN
Boston   Sydney

First published in 1981

GEORGE ALLEN & UNWIN LTD
40 Museum Street, London WC1A 1LU

© G. C. Brown and A. E. Mussett, 1981

British Library Cataloguing in Publication Data
Brown, G C
  The inaccessible earth.
  1. Earth – Internal structure
  I. Title   II. Mussett, A E
  551.1′1   QE509

  ISBN 0–04–550027–4
  ISBN 0–04–550028–2 Pbk

Set in 10 on 13 point Times by Servis Filmsetting Ltd,
Manchester and printed and bound in Great Britain by
William Clowes (Beccles) Limited, Beccles and London

Preface

This book is about the Earth, its formation, evolution and, particularly, its present internal state and composition. It is here that the two great branches of the modern Earth sciences, geophysics and geochemistry, meet, though they are rarely combined in undergraduate texts to give a unified view of the Earth's interior.

Our understanding of the deep Earth has changed prodigiously in the past two decades and, although the future no doubt will see major changes in our understanding, there are signs that most advances will be a steady consolidation of present ideas, in a Kuhnian 'mopping up' phase. The problem in writing this book, then, has been one of amalgamating recent evidence, not just from geophysics and geochemistry, but also embracing parts of astronomy, meteoritics and so on. It is this breadth of input that makes the subject such a fascinating one but, because the available information is scattered through a wide range of books and papers – much of it is in specialist form – it is seldom read by the undergraduate.

Because our intended audience is generalist rather than specialist we have concentrated on getting clear the underlying physical and chemical principles, using mathematics only sparingly. Boxes in the text are used for particularly difficult concepts which it seemed important to develop for the more specialist reader. Conversely, for the less well-versed reader, some important basic groundwork, common to most undergraduate courses, is summarised in Notes at the end of the book. A 'Further reading' list appears at the end of each chapter. This includes non-specialist reviews or books or, when these are not available, recent papers that provide an entry into the literature. Those wanting to go more deeply into the subject can use the references towards the end of the book which include all but the oldest of the works cited. Two diagrams, in the front and back flaps, are included as overall summaries for easy reference.

If any of the views in the book do not find favour with our academic peers, then the responsibility is ours alone. But if the book achieves its aim of providing an accessible, generalist, undergraduate book, then our warmest thanks must go to all the friends and colleagues who helped with reviews and comments on draft chapters; in particular, Bill Fyfe, Ian Gass, Peter Harris, Aftab Khan, Richard Cooper, Peter Dagley, Peter Francis, Bob McConnell, Currie Palmer, Richard Thorpe, Rod Wilson and Brian Windley. We also thank our two typists, Pauline Lybert and Sue Hartnett, who prepared reams of typescript, often under great pressure from impatient authors. Finally, G.C.B. is deeply grateful to Joan and the children for their patience and encouragement through the many long, uncommunicative hours during the preparation of this book; A.E.M. is merely thankful he has no wife.

Contents

Postscript: The state of ignorance 206

1 Introduction

1.1　Aims and objectives

This book is about the interior of the Earth. It discusses the answers to such questions as: Of what is the Earth made? How does it behave? In other words, what **chemical elements** are present in different parts, and how are they combined to form the compounds that are built up into minerals and rocks? Also, what are their **physical properties:** are they liquid or solid, can they conduct electricity, and so on? The interior of the Earth is known to be a dynamic system because of its mobile surface features, expressed through the motions of continents, for example, and this raises two other important questions: What forces and what processes are operating, and how has the Earth changed since our planet was formed about 4600 million years (Ma) ago?

We should like answers to these and many other questions, but the difficulty is that most of the Earth's interior is inaccessible to direct sampling or observation. It is true that rocks now at the surface, brought up in diamond-bearing kimberlite pipes and other volcanoes, have risen from depths of as much as 200 km, but, the deeper their source regions, the more likely it is that changes have occurred during their ascent to the surface. And even 200 km is but a small fraction of the Earth's radius (6370 km); so, valuable as such information is, it is far from sufficient to reveal the gross constitution of the Earth's interior.

There is no simple solution to this problem of inaccessibility. We have, instead, to combine information of many kinds, and it is the diversity of sources which makes the study so fascinating: astronomy, astrophysics, nuclear theory, planetary physics, as well as geophysics, geochemistry and geology all play a part. Often, each piece of information only narrows the possibilities, but if enough constraints are applied together we may arrive at a close approximation to the true constitution of the Earth. For example, the core is thought to be chiefly iron, for only iron is sufficiently dense, is an electrical conductor needed to produce the Earth's magnetic field, and is likely to be sufficiently abundant. No other material can satisfy all these requirements. But, before these techniques are discussed, our first object is to give the reader an impression of the approach and general conclusions of the book.

1.2　'The inaccessible Earth': an outline

To simplify matters, the authors have chosen the following plan which gives a logical development, but which does not necessarily adhere to the strict chronological evolution of knowledge.

The first contribution comes from seismology, the most useful single discipline concerning the inside of the Earth. Earthquakes generate waves that may penetrate deep into the interior before eventually emerging far from their source. There are several sorts of seismic wave, and the paths that the different kinds of wave follow depend upon the way the seismic wave velocities vary within the Earth –　1 which, in turn, depend upon the physical properties of the interior; thus, by painstaking analysis of seismic records obtained at many recording stations set up around the world, it is possible to deduce how the seismic velocity varies within the Earth. The picture that emerges is closely that of a concentrically-layered Earth in which the most striking division is the change, about half-way to the centre, from a solid (the **mantle**) to a liquid (the **core**). However, there are many other discontinuities or rapid changes, some of which are shown in Figure 1.1, while a fuller summary is provided inside the front cover of the book.

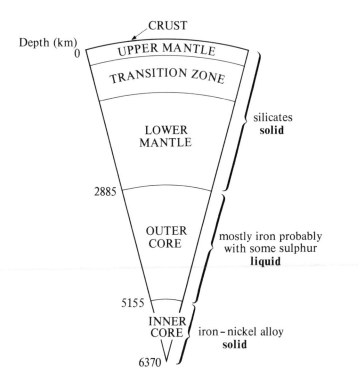

Figure 1.1 Sector of the Earth. Only the major features are shown. A fuller summary is given inside the front cover.

To find out whether these changes are due to changes of composition, temperature or other parameters, we turn to density, which is the most useful quantity that can be determined with reasonable precision. Density is deduced by combining information from seismology with a knowledge of the mass of the Earth, deduced from its gravitational attraction, and its moment of inertia, determined from the movement, or precession, of its axis of rotation. The resulting density variation with depth inside the Earth is not uniquely determined, but it is known to within fairly close limits at most depths, so that next we can ask what material will match the density at any specified depth.

Obviously, a vast number of substances can have a particular density, so the purely physical evidence used so far has to be reinforced by other types of information that constrain the possible chemistry of the Earth. For example, current theories about the way in which the Solar System formed suggest that the Sun and planets condensed from a cloud, or **nebula**, of gas and dust. **Meteorites** play a prominent role here, for they are believed to represent early stages in the process of planet formation. In particular, a few are thought to be remarkably close in composition to the original nebula, thereby providing an approximate bulk composition for the Earth.

It is well known from seismic studies that the Earth is not a homogeneous body, but that the chemical elements have been segregated, or **differentiated**, into the layers mentioned earlier. There is some debate about the timing of this segregation: the consensus view is that some segregation occurred while the Earth was being accreted and that some has occurred subsequently. In recent decades, it has been demonstrated from crustal studies that some of these processes have operated throughout the Earth's history, albeit with a declining vigour as time has progressed. Given the constraints of temperature and also pressure variation in the Earth, the simple rules of geochemistry can be used to predict the combinations of chemical elements that can coexist stably as minerals at various depths. We are now in a position to apply these considerations to each region of the Earth in turn, together with any other relevant information.

The first region considered is the core. It is composed chiefly of iron, and the problems remaining are: to determine what small quantities of other elements must be present to bring the density into agreement with the calculated density variation throughout the core; to explain the existence of the solid inner core; and to account for the source of energy needed to produce the Earth's magnetic field.

Moving outwards, the mantle is thought to be far more complex. This is because it contains more elements in major amounts than does the core, and because the various minerals formed from these elements change their *crystalline* form in response to the variations in pressure and temperature. Also, in the upper part of the mantle, the material is near to its melting temperature, which is important for two reasons. First, it gives this part of the mantle a greater plasticity when subjected to long-continued stresses, and secondly, a molten or liquid fraction may be produced which, on rising to the surface, leads to igneous activity. The plastic property of the mantle allows the semi-rigid outermost part of the Earth, the **lithosphere**, to be moved both vertically and horizontally, by thermal convection, leading to mountain building and continental drift. Were it not for these movements, the Earth's surface would be almost featureless.

The crust, the uppermost part of the lithosphere, is the most complex region of the Earth, because the cyclic phenomena of sedimentation, metamorphism and igneous activity which process and reprocess its materials can lead to the extreme differentiation of the chemical elements. Principally through the techniques of stratigraphy and petrology, we shall be concerned with large-scale features of the continental crust and the processes that exchange heat and material between the crust and upper mantle. The continental crust has been evolving irreversibly throughout geological time and also has probably grown in bulk at the expense of the upper mantle. Because of crustal growth and thickening, and because of the decreasing rate of heat production by radioactive decay, mountain building and other major tectonic processes may have changed in style as well as intensity during the 4600 Ma of the Earth's history. Associated with the evolution of the Earth have been changes in the composition of the atmosphere, the types of sedimentary rocks deposited, and life itself.

The remainder of this chapter concerns the history of our knowledge of the Earth and its interior. Throughout the history of science, an understanding of the Earth's interior has had to await many of the most recent advances in physics, chemistry and astronomy, as well as geology. But it is noteworthy that, throughout scientific history, Earth science has been no fringe subject. Often, it was at the forefront of science, giving rise to controversies, such as those between the proponents of spontaneous creation and natural evolution, or between the literal adherents of the biblical age of the Earth and those who studied radioactivity, or slow geological processes.

It is customary in this kind of review to refer either to the Chinese or to the Greeks or even, if erudition is excessive, to more remote civilisations. In this case, it is to the Greeks. In the period of 600–200 BC, their wealth of intellectual enquiry and speculation was amazing, and it had a major formative influence on our own ideas, particularly through the writings of Aristotle. Geological matters formed only a minor part of their interests, but they realised that land could become submerged, or form out of the sea. In addition, they understood that fossils represent organisms buried by past seas. On a larger scale, they knew that the Earth is a sphere and devised means of measuring its radius to within a few per cent. They also believed in a central fire, a belief that was to recur throughout history. One philosopher, Aristarchus (310–250 BC) even put forward the heliocentric (i.e. Sun-centred) theory of the planetary motions, but the idea did not find favour, and the geocentric (Earth-centred) theory was almost universally accepted until the 16th century AD.

Greek scientific reasoning had one serious weakness, which was its heavy reliance on elegant solutions that were deduced by reason, with little recourse to observation and experiment. For example, Socrates expelled a pupil from his logic class for suggesting that the best way to calculate the number of teeth a horse has was to open the horse's mouth and count them. As a result, they were unable to take many of their ideas beyond speculation.

The Greek body of thought was absorbed by the Romans, but with the fall of Rome in the 5th century AD it was almost lost, being preserved only in the Byzantine Empire. However, much of this knowledge percolated through the Islamic world and was eventually transmitted to the West, together with other knowledge, such as the decimal system, which originated in India. At first, mediaeval Europe regarded this pagan knowledge with suspicion, but gradually it was reconciled with Christian thought by the scholastic philosophers, prominent among whom (by girth as well as intellect) was St Thomas Aquinas (AD 1225–74). As a result, the works of Aristotle came to have an authority second only to that of the scriptures. This influx of knowledge into the West paved the way for the great upsurge of science, discovery and general enquiry that characterised the Renaissance.

But, before science as we know it could blossom, the reliance on authority to settle points at issue had to be challenged; it took a long time to accept that observation can overrule the stated opinions of eminent persons (practice has not quite caught up with this precept!). It is hard for us now to realise the true nature of this revolution in thought: an extreme act of faith must have been required to induce

4

belief in rational investigation of a world supposedly populated with witches, hippogriffs, and a mineralogy that included jewels in toads' heads. But, gradually, observation became accepted as the final arbiter.

Copernicus (AD 1473–1553), a Pole, was one of many dissatisfied with the geocentric theory that had the Earth stationary at the centre of the universe. The version of this theory then prevailing was due to Ptolemy who lived in Egypt from AD 90 to AD 168. To account for the fact that the planets occasionally appear to reverse their motion across the sky (see Fig. 1.2), Ptolemy had proposed that they moved in epicycles. Copernicus probably took his idea of a heliocentric system from Aristarchus, for he admitted as much in a paragraph which later he deleted. To avoid criticism from the Church for demoting the Earth from the centre of the universe, a friend, who had been entrusted with publishing Copernicus' ideas, inserted a preface stating that the theory was merely a convenient method for simplifying calculations!

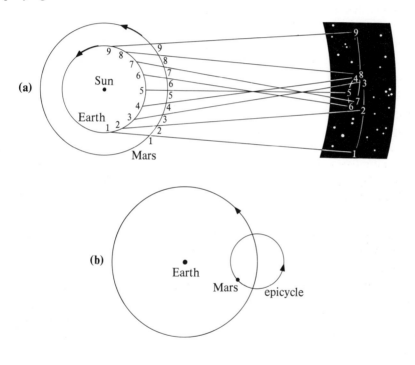

Figure 1.2 Illustration of epicycles. (a) Because planets take different times to make one revolution about the Sun, a planet, viewed from the Earth, appears to form a loop in its motion. The numbers show the positions of the Earth and Mars at the same successive instants. (Redrawn from *Structure and change* by G. S. Christiansen and P. M. Garrett. © 1960 W. H. Freeman & Co.) (b) To explain such loops, the Ptolemaic system, which puts the Earth at the centre of the universe, made Mars move in a small circle (epicycle) which itself travels around the Earth. (Redrawn from *New horizons in astronomy*, 2nd edn, J. C. Brandt and S. P. Maran (eds). © 1976 W. H. Freeman & Co.)

Tycho Brahé, a Dane who lived from 1546 to 1601, did not accept Copernicus' theory but, unlike Copernicus, he was a great observer and he produced greatly improved data that paved the way for Kepler (1571–1630), his German assistant, to eliminate epicycles by describing the rules of elliptic orbits. Galileo (1564–1642) also helped to disprove the Ptolemaic system; as a result he incurred the displeasure of the Italian Inquisition, who forced him to recant. On his death, he was denied a monument in the hope that he and his works would be forgotten, a vain hope, for the year of this death was that of Newton's birth in England.

Sir Isaac Newton (1642–1727) set the seal on these international advances with his theory of gravitation, and showed that if the force of attraction between two bodies were proportional to the product of their masses (m) and inversely proportional to

5

the square of their separation (*d*), i.e.

$$F \propto \frac{m_1 m_2}{d^2}$$

elliptic orbits followed naturally. Newton was partly inspired to his theory by the work of William Gilbert, physician to Queen Elizabeth I, who had written the first geophysical treatise, a perceptive account of the Earth's magnetic field. Newton was unable to test his theory of gravitation directly by measuring the very small force between two masses in the laboratory. However, he showed theoretically that, under the action of gravitational and centrifugal forces, the Earth ought to have an equatorial bulge. Attempts by the French to measure the shape of the Earth seemed at first to show an elongation along its axis. To settle the matter, the French Academy organised expeditions to different latitudes in the decade 1735–45, and these showed conclusively that a bulge exists.

Newton's formulation of the theory of gravitation and of the laws of motion marks a watershed in the development of modern science from mediaeval belief and alchemy. It, therefore, comes as something of a shock to learn that Newton regarded his many scientific discoveries as a minor part of his work, and most of his 20 million words output concerned theology and such almost inconceivable topics as the topology of hell.

Progress in geology was not so dramatic. As a descriptive science, many of its early practitioners could only establish laws by the painstaking accumulation of many observations, and not by the elegant mathematical breakthroughs of astronomy and physics. However, systematic geology only began with the observation of strata which had practical applications in mining and engineering. Gradually, the idea of water-lain sediments became accepted, but fossils within them caused a lot of trouble. In mediaeval and later times, fossils included not only animal and plant remains but also curiously shaped minerals. It was also thought that fossils could form spontaneously within a rock. However, Leonardo da Vinci (1452–1519) understood clearly that fossils found deep within hills were the remains of long-dead creatures, and could not be attributed to the biblical deluge. Steno (1638–1686) documented certain of the rules of stratigraphy; for example, the law of superposition which states that the lowest layer in a succession must be the oldest, and so he paved the way for geological mapping by relative ages. William Smith, a surveyor and canal builder, applied these ideas, and also used fossils to correlate strata from different places. In 1815, he produced the first (hand-coloured) geological map of Britain, and in the following year published a book, *Strata identified by organised fossils*.

It has become a well established principle of scientific method, known as Occam's razor, not to use two hypotheses in an explanation if one can be found to serve the purpose. But some geologists carried economy of hypotheses to fanatical lengths and claimed an all-embracing validity for their theories, forgetting the fact that, outside the carefully controlled conditions of the laboratory, many unrelated processes may be operating simultaneously. For example, the importance of water in forming rocks was seized upon by the Neptunists to account for nothing less than the formation of all the solid Earth. The theory was championed by Abraham Werner (1750–1817), Professor at Freiberg University, who believed that virtually all rocks were formed either by precipitation or crystallisation in a universal sea.

The theory was simple, but much of it was arbitrary and beyond test. However, the sheer volume of water involved was a problem, for not only did it have to disappear after the initial formation but it had to return for succeeding epochs of deposition (the biblical deluge was only the last of many such epochs).

Neptunism was opposed by the Plutonists, whose cause was expressed by Sir James Hutton of Edinburgh (1726–1797) in his book *Theory of the Earth*, one of the most important geological texts ever written. The Plutonists did not dispute that sedimentary deposition occurred, but they did not believe that this could lead to hard rocks without the action of heat. Their theory was less simple than that of the Neptunists, but it did offer a more dynamic Earth in which rock formations could be lifted and tilted and so oppose the levelling effects of erosion. Disagreement between Plutonists and Neptunists sometimes reached a vigour not encountered in present-day controversies; in Edinburgh, a play written by an ardent Huttonian was hissed off the stage by an audience of Neptunists!

The Neptunist–Plutonist controversy merged within a generation into another: Catastrophism versus Uniformitarianism. This issue particularly concerned time: nowhere has the Bible been more restrictive to geology than concerning the age of the Earth. The computation in 1664 of Archbishop Ussher, a contemporary of Newton, that the Earth was created at 9.00 a.m., 26 October 4004 BC is well known. But this did not suit some geologists who had begun to appreciate that slow geological processes could move mountains given sufficient time.

That geological change had occurred could no longer be denied. Catastrophism accounted for most of this change by a series of immense upheavals due to supernatural forces, while between these 'catastrophes' geological processes resumed normal operation, but produced only relatively small changes. The advantage of this theory was that a great deal of change could be accomplished in a very short time and so it could be reconciled with the biblical time scale. It had the drawback, from the point of view of science, that it was beyond the reach of rational enquiry. By contrast, the theory of Uniformitarianism emphasised the continuity of geological processes. Hutton had observed geological unconformities and recognised that sedimentary successions could be built up, tilted and then eroded during a very long period of time, after which another cycle of deposition occurred. He enunciated the principle that processes in the past were the same as those observable today: 'The present is the key to the past'. This offered a rational means of explaining how rocks are formed, in terms of comprehensible forces that carry implications for the conditions existing during formation. For instance, the presence of marine fossils shows that the land was once below sea level. A picture of some past environment can be built up and tested for self-consistency: it is therefore a very powerful method of investigation.

The ideas of Uniformitarianism were developed and clarified by Playfair (1748–1819) and by Sir Charles Lyell of Edinburgh (1797–1875). Lyell, in his book *Principles of geology*, published in 1833, showed that existing processes could account for observed geological changes, but vast periods of time were required. Lyell's arguments were persuasive, and the ridiculously short biblical time scale gave way to immense tracts of time. But the pendulum swung to the other extreme: Hutton's cautious statement that he could see no evidence of a beginning or an end to geological processes became, for some of this disciples, a world without end, with geological cycles of erosion and mountain building continuing indefinitely. However, opposition to such an infinite time scale was soon to come.

Physicists had been establishing the laws of thermodynamics, which demonstrated that processes could not continue indefinitely: the mechanism had to run down as the available energy was expended. Prominent among them was Lord Kelvin who, in the middle of the last century, calculated the time for the Earth to cool from a molten ball to its present temperature, which he estimated from the temperature gradient in mines, and the existence of volcanoes. He deduced a value of only about 100 Ma. This figure dismayed not only the geologists, but also Darwin, whose theory of evolution (published in 1859) owed much to Lyell's ideas, and was only possible if much longer intervals of time were available for evolution to bring about its changes. We know now that the Earth gets much of its heat from radioactivity, but at that time Kelvin's case seemed unassailable; physical laws allied to mathematics seemed more than a match for semi-quantitative estimates of time based upon rates of geological or biological processes.

More and more geologists, over a period, came to accept the 100 Ma estimate, until it became almost a dogma for them, but the physicists were refining their estimate of the age of the Earth, usually downwards, and in some cases to as little as 20 Ma; this the geologists found progressively more constricting and it forced them to refine their calculations – based upon processes such as rates of sedimentation and the total thickness of sediments – until they were sufficiently confident to challenge the physicists. But then, at the end of the last century, radioactivity was discovered and it was soon realised that the physicists' estimate of cooling time should be regarded as more like a minimum value for the age of the Earth, rather than a maximum. So another fruitful controversy ended, with the acceptance of a time scale of hundreds of millions of years, which opened up new possibilities not only in geology but also in the life sciences and cosmology. (The current estimate of the age of the Earth – really of the formation of the Solar System – is about 4600 Ma. The way this is deduced is explained in Note 6 at the rear of this book.)

Buffon was the first to attempt to deduce the internal constitution of the Earth in a realistic manner. He believed that a molten interior is needed to allow the equatorial bulge to form due to rotation, and in 1776 published a theory whereby a molten Earth was formed when a comet collided with the Sun and knocked out material. He suggested that the subsequent evolution of the Earth required the more refractory materials to solidify first, leaving the volatile ones to form the oceans; later, the continents separated out and, finally, Man emerged. Buffon performed crude experiments to measure the cooling rates of masses of various materials. By extrapolating the results to a mass the size of the Earth, he deduced a total cooling time of about 75 000 years, thereby foreshadowing Kelvin's method for estimating the age of the Earth.

In 1828, Cordier made measurements of the temperatures in mines and obtained a value for the geothermal gradient near the surface of the Earth of $30°C\,km^{-1}$, in remarkable agreement with modern values. This was additional support for a hot interior of the Earth whose heat was attributed (since radioactivity was not known at that time) to an initially molten state.

A further inference from the cooling Earth model was that the solidified crust would buckle into mountains as the interior cooled and contracted. However, a new explanation of mountain belts was required once radioactive heat production was discovered around the turn of the century, for this gives the Earth a much longer active span, with little if any cooling. Nevertheless, the early geologists were right to see internal heat as the major source of energy which elevates the Earth's surface and which acts in opposition to the destructive forces of erosion powered by solar energy.

Earthquakes, naturally, had long attracted attention. Explanations offered in classical times involved movement of internal water supporting the surface of the Earth, fire bursting out of the interior, or collapse of caverns. In mediaeval times, these mechanical, if vague, ideas were replaced by animalistic explanations; for example, the restless movement of a giant serpent in the depths of the sea was proposed seriously. Modern ideas began with Hooke, who recognised in 1705 that earthquakes are associated with land movements, and Mallet (1810–81), in the middle of the last century, realised that most earthquake damage results, not from the gross movement of the land itself, but from waves that are generated by the movement and spread out from their source. With great perspicacity, Mallet suggested that by timing the arrival of seismic waves which have travelled right through the Earth it should be possible to increase our knowledge of the Earth's interior. He was right, and seismology has proved to be a major tool for exploring the inside of the Earth.

The existence of earthquakes, volcanism and the growing body of evidence for geologically recent changes in sea level made it increasingly clear that the Earth had not only changed in the past but is still a dynamic planet. Yet, what forces were operating to cause vertical movements and how did the Earth respond to them? One clue came from an unexpected quarter. An attempt by Bouguer in 1735–45, to 'weigh the Earth' by comparing the gravitational pulls of the Earth and the Andes, showed that the mountains apparently had considerably less mass than expected for their volume. Later, the effect was found elsewhere, particularly in the Himalayas during surveys of northern India by Sir George Everest and others. It was suggested that the deficit in mass is due to the rocks beneath the mountain being lighter than in adjacent areas. But how could this come about? As the effect was too common to be assigned to chance, there had to be some mechanism that would bring it about naturally. In 1855, Pratt and Airy each published the concept of **isostasy**, the idea that surface rocks float on a layer of denser but yielding rock, and stand higher where the lighter rocks are thicker.

This was only a partial explanation of mountain building, for it did not explain how the lighter rocks came to be accumulated into thick masses in the first place, a problem that has been solved only with the advent of plate tectonics in recent times. Plate tectonics is a development of the earlier idea of continental drift which was first put forward in a coherent form in 1910, by Alfred Wegener, a German meteorologist. He argued on the basis of different kinds of evidence of matching, that all the continents had been part of a single supercontinent several hundred million years ago, which since had broken up. Acceptance of continental drift only became widespread with the blossoming of palaeomagnetism, in the 1950s, which provided independent evidence for the movement of continents.

About the same time, techniques were being developed to investigate the two- **9** thirds of the Earth's surface which are under water, with the recognition that the ocean floors are quite different from the continents. This difference was explained by the theory of plate tectonics which, unlike continental drift, embraced the whole surface of the globe, and was formulated in the later 1960s. It states that the surface of the Earth is divided into a few huge pieces, or plates, which are rigid to a good approximation, so that all forms of tectonic activity are largely confined to the plate boundaries where there is relative motion of the plates. The phenomena observed – earthquakes, volcanism, mountain building, etc. – depend upon whether plates are moving towards or away from each other, and whether the parts in contact are continental or oceanic.

All these large-scale phenomena – isostasy, volcanism, plate tectonics – presumably are due to processes occurring deep in the Earth and so they have helped to focus interest on the interior. And that, of course, is the subject of this book, to be discussed in the chapters that follow.

Further reading

General books:
Adams (1938); Haber (1959); Gillespie (1951): history of geological reasoning.
Burchfield (1975): The early controversy between geologists and geophysicists about the age of the Earth.
Mather & Mason (1939): Source book in geology.
Runcorn (1967): *Dictionary of geophysics.*

2 The contribution of seismology

2.1   Seismic waves

Seismology can mean the study of earthquakes, but it also means the study of the interior of the Earth by means of seismic waves which can penetrate right through it. Much of the information comes from the times it takes the waves to travel different distances.

Seismic waves can be generated by any disturbance of the ground, but to be detectable on the other side of the Earth only earthquakes and nuclear explosions are large enough **sources** of waves. The waves produced are of four kinds, as illustrated in Figure 2.1. They can be divided into body and surface waves: body waves travel within the Earth, while surface waves are largely confined to near the surface. The body waves in turn are of two kinds. **P-waves** are longitudinal waves, which means that the individual particles of the material in which the waves are travelling oscillate back and forth in the direction of propagation (Fig. 2.1a), and they are simply sound waves within the Earth. **S-waves**, on the other hand, have their oscillations transverse to the direction of propagation (Fig. 2.1b).

A wave is the propagation of a strain through a material (Fig. 2.1). If any small volume within an elastic medium is distorted or **strained** in some way, a **stress** is developed which tends to restore the material to its undisturbed state. The ratio of stress to strain is the **elastic modulus** of the material. The speed of propagation increases with the modulus but decreases with the density of the material. The general expression for the wave velocity is

$$\text{wave velocity} = \sqrt{\frac{\text{elastic modulus}}{\text{density}}} \tag{2.1}$$

Because a material can be strained in different ways, and so has more than one elastic modulus, it can have more than one wave velocity. The expressions for the velocities of the two seismic body waves are:

$$V_p = \sqrt{\frac{K + \frac{4}{3}\mu}{\rho}} \qquad\qquad V_s = \sqrt{\frac{\mu}{\rho}} \tag{2.2}$$

where ρ is the density, K is the **bulk** or **compressibility modulus** which is a measure of the stress (i.e. force/unit area) needed to compress a material to a smaller volume, and μ is the **rigidity** or **shear modulus** which is a measure of the stress needed to change the shape of the material. Examination of Figure 2.1 shows that a P-wave both compresses and changes the shape of the material, and hence V_p depends upon both K and μ. S-waves, on the other hand, merely change the shape of the material.

Examination of the two formulae of Equations (2.2) shows that V_p must always be greater than V_s, and so P-waves from an earthquake will always arrive before S-waves at a seismograph recording station (see Fig. 2.2). Before their nature was known, these arrivals were simply called primary and secondary waves, and the

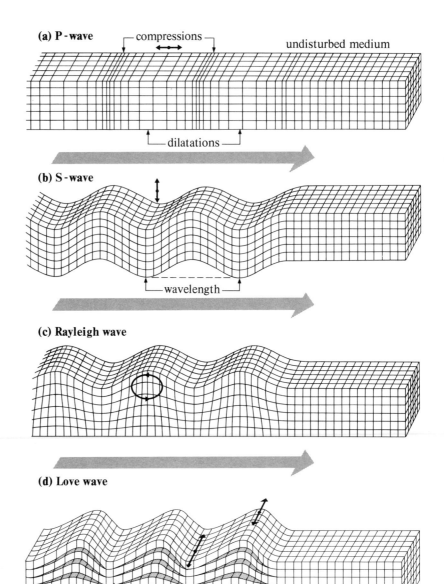

(a) P-wave

- compressions -

undisturbed medium

- dilatations -

(b) S-wave

- wavelength -

(c) Rayleigh wave

(d) Love wave

Figure 2.1 Four types of seismic wave. (a) In P-waves, particles oscillate to and fro along the direction of propagation of the wave; (b) in S-waves, particle motion is transverse. (c) The particle motion of Rayleigh waves is more complex, but near the surface is a backwards vertical ellipse; (d) in Love waves, it is transverse and horizontal. In both Rayleigh and Love waves, particle motion decreases with depth from the surface. (Based on a figure in *Nuclear explosions and earthquakes: the parted veil* by B. A. Bolt. © 1976 W. H. Freeman & Co.)

initials P and S have remained. However, it may help to think of them as pressure and shear waves. A second deduction we can make from the formulae is that S-waves cannot propagate in a material such as a liquid, which cannot resist a change of shape and so has no rigidity (i.e. $\mu = 0$). This information is important because we shall use it to deduce that part of the Earth's core is liquid (Section 2.3).

The other two types of wave shown in Figure 2.1 are surface waves, for they can be generated only if a surface – generally the surface of the Earth – is present, and their amplitude decreases rapidly below the surface. For this reason, and because they are slower than P- and S-waves, they played little part in the early development of

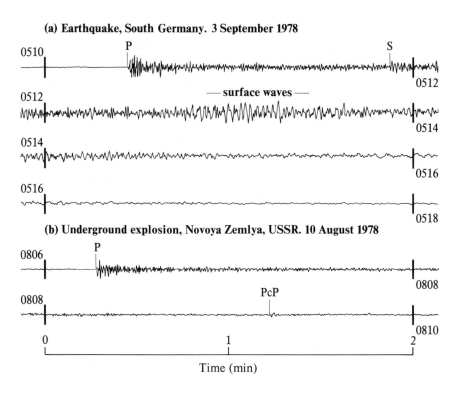

(a) Earthquake, South Germany. 3 September 1978

Figure 2.2 Seismic recordings. The upper record is for an earthquake only 8° from the recording station and shows P-, S- and surface waves. The lower record is of a Russian explosion which, being 31° from the recording station, allows the PcP arrival to be seen clearly. Both were recorded. at Leicester University's seismic station (CWF) in Charnwood Forest, U.K., and were provided by courtesy of Dr M. A. Khan. (The PcP arrival is defined in Fig. 2.5 and the angles are epicentral, subtended by source and recording station at the centre of the Earth.)

global seismology. This is no longer true and they will be considered further in Section 2.4.

The major object of this chapter is to deduce V_p and V_s at all depths in the Earth and so show that the interior can be divided into distinct concentric zones, separated by discontinuous, or rapid, changes in velocity. A further purpose of this chapter is to establish other data depending upon the moduli and density, so that in Chapter 3 we can quantify variations in density, for density provides a much better clue to the composition at depth than do the seismic velocities.

2.2 Deduction of P- and S-wave velocities at depth

We need first to understand the rules that govern wave propagation within the Earth. A 'point' source of seismic waves, such as an explosion or an earthquake, will generate waves that spread out spherically, but when they encounter a region with different density or elastic properties – and hence different seismic velocity – these spherical wavefronts will be distorted. This is closely analogous with the behaviour of light, and, as with light, it is usually more convenient to think in terms of rays rather than wavefronts, with a ray being the path followed by a small section of wavefront, as illustrated by Figure 2.3. The section of wavefront AB moves to CD, where AC and BD are in proportion to V_1 and V_2, the seismic velocities of the two materials, so we can write:

$$\frac{\sin i_1}{\sin i_2} = \frac{AC/BC}{BD/BC} = \frac{AC}{BD} = \frac{V_1}{V_2} \tag{2.3}$$

13

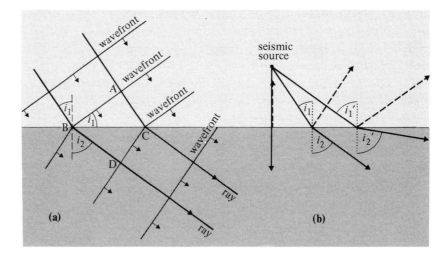

Figure 2.3 Refraction and reflection of seismic waves. (a) Parallel waves enter a material of higher wave velocity. This causes the wavefronts – equally spaced in time – to slow, so that the time to travel from B to D equals that from A to C. (b) Selected rays diverge from a point source and are refracted at the interface. All obey $\sin i_1 / \sin i_2 = V_1 / V_2$ where V_1 and V_2 are the velocities in the upper and lower layers, respectively. In addition, there are the reflected rays, so that not all the energy is transmitted through the interface.

This is the basic rule governing refraction at an interface between two materials and is analogous to Snell's law in optics.

In the Earth, most changes of velocity takes place continuously, rather than abruptly at definite interfaces, and this we can approximate by treating such regions as a series of thin concentric shells (Fig. 2.4) and applying Equation (2.3) at each interface. In this way it would be possible, provided we knew the velocity at all depths, to trace different rays through the Earth until they re-emerge at the surface, and also to calculate how long this would take (Fig. 2.4c). Unfortunately, our problem is the inverse: the velocity–depth profile is what we should like to know but, confined as we are to the surface of our planet, all we can observe is the *time* for seismic waves to arrive at different places around the Earth.

Figure 2.4 Schematic diagram of rays in a concentrically layered Earth. (a) Changes in velocity can be approximated by many layers with small increments in velocity. (b) An increase of the velocity downwards causes rays to bend back to the surface, after having reached some maximum depth. (c) Diagram showing the corresponding time–distance or travel-time curve. Note that distances are measured by the angle subtended at the centre of the Earth, the epicentral angle, Δ.

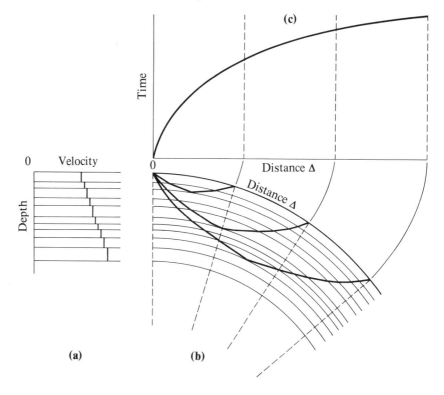

The problem of **inversion** – the deduction of what lies below the surface from measurements taken on or above the surface – is a common one in geophysics. It is generally much more difficult to solve than the direct case, but for this seismic problem, mathematical expressions can be derived (Note 1) that permit the inversion of the **travel-time curve** (Fig. 2.4c) to give the velocity–depth profile. But, first, we shall look at the way accurate and detailed time-distance graphs have been built up over the years from the careful analysis of many earthquake records.

The process begins when a large earthquake occurs and waves from it are received at seismic stations scattered around the world. The next step is to identify the various arrivals at the different stations for, in addition to the P-, S- and surface waves so far mentioned, there are many other ray paths involving reflection and/or refraction within the Earth, some of which are shown in Figure 2.5a. (These, however, can be important for improving our knowledge of variations of velocity within the Earth.)

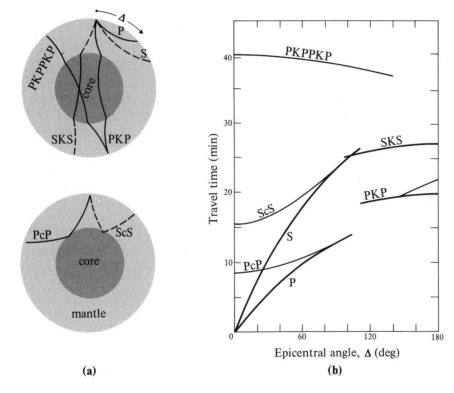

(a)　　　　　　　　　　　　**(b)**

Figure 2.5 Selected rays and corresponding travel-time curves. (a) In the mantle, where the velocity, in general, increases with depth, the rays bend away from the vertical. Those reaching the core may either be reflected back into the mantle, or refracted into the core, which has a lower velocity. Capital letters denote both the type of wave and the part of the Earth through which it is travelling, K denoting a P-wave in the core; c indicates a reflection at the core. Rays can experience many more reflections and refractions than shown. (b) Corresponding travel times. In practice, of course, (a) is deduced from (b) and not vice versa. (From Richter 1958 and Jeffreys 1962.)

As yet, neither the location nor time of the earthquake are known. These can be estimated by comparing records from the different stations, for those closest to the earthquake will record its arrival earliest and with the largest amplitude. In this way, rough travel-time curves for P-and S-waves could be constructed. Once this is done, we have a better method for estimating time and distance of earthquakes, using the difference in arrival time of the P-and S-waves. Figure 2.5b shows how this interval is related to the distance of the seismic station from the earthquake (expressed as the epicentral angle Δ). Thus, the time and distance of the earthquake can be read off. One station is not enough to determine the location of the earthquake, only its distance, but by using several stations it is possible to pin-point its position, as

15

illustrated in Figure 2.6. If the travel-time curve (Fig. 2.5) is not accurate, there will be several intersections of the curve instead of a single unique one, and so the curve can be adjusted to minimise the discrepancies.

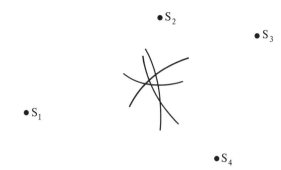

Figure 2.6 Location of an earthquake. The interval between the arrival of P- and S-waves at a seismic station (see Fig. 2.2) allows the distance of the earthquake to be deduced from travel-time curves (Fig. 2.5b). If it has been received at several stations, S_1 to S_4, its position can be found by drawing arcs as shown. Lack of a single point of intersection shows error in the travel-time curves and can be used to correct them.

A single earthquake only gives values on the travel-time curve corresponding to the distances of the various stations recording it. To build up almost continuous curves, as shown in Figure 2.5, requires the recording, over many years, of earthquakes occurring all over the world. This synthesis of a continuous travel-time curve from many earthquakes is only possible because the Earth is spherically symmetrical to a high degree, after allowing for the equatorial bulge due to rotation (Section 3.3.1). Nowadays, travel times are know to an accuracy of a few seconds, in a total time that may be as much as 20 minutes. Most of the remaining discrepancies are due to inhomogeneities within the Earth (Section 8.9).

The principles used to invert the travel-time curves into a velocity–depth profile for P- and S-waves are outlined in Note 1 at the rear of the book. Nowadays, this method is supplemented by what is a sophisticated method of trial and error. A velocity–depth curve is guessed and travel times are computed for various epicentral angles and compared with observed times. The velocity–depth curve is then adjusted to improve the agreement, in a continuing iterative process. Travel-time curves for more complex paths, such as PcP and PKPPKP (Fig. 2.5) are used as additional tests of the selected velocity–depth curves.

2.3   Seismic velocities and the structure of the Earth

Figure 2.7 shows a recent profile, together with some alternatives within the core, where there is considerable uncertainty. The outstanding feature is the abrupt decrease in P-wave velocity about 2900 km down, accompanied by the disappearance of S-waves. It is this **discontinuity** which, by definition, divides the mantle from the core. We infer, from the absence of S-waves, that the outer core is liquid because, as we saw in Section 2.1, S-waves cannot propagate in a liquid.

The abrupt decrease in V_p at the core–mantle boundary causes rays entering the core to be refracted *towards* the vertical, and, as a result, P-waves cannot reach the Earth's surface within a zone between 103° and 142° from the earthquake, forming a **shadow zone**, as shown in Figure 2.8. However, it is not a perfect shadow zone, and because of this Miss Lehmann postulated, in 1936, that there is an inner core in which V_p increases by a considerable amount, causing rays to deviate away from the

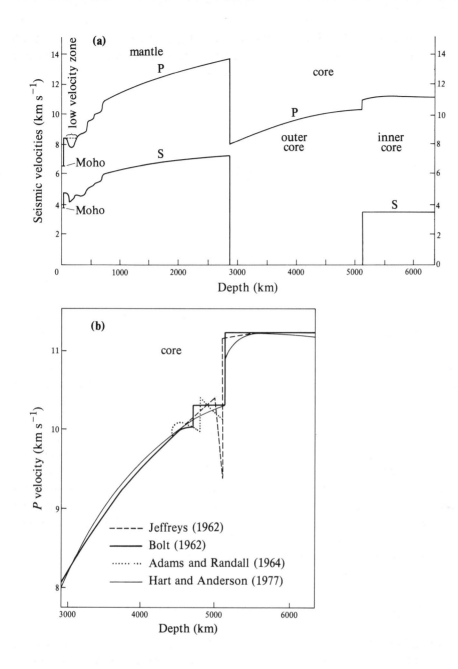

Figure 2.7 Seismic velocity versus depth profiles. (a) A recent profile for the whole Earth (Hart *et al.* 1977). Future revisions are likely to be small, except perhaps for around the inner core–outer core transition. (b) Some suggested core profiles, illustrating the degree of uncertainty about the inner core–outer core transition. (Jeffreys (1962) is reprinted from *The Earth*, 4th edn, by Sir Harold Jeffreys, Cambridge University Press: Bolt (1962) is reprinted by permission of *Nature* **196**, 22–4. © 1962 Macmillan Journals Ltd: Adams and Randall (1964) is reprinted by permission of the Seismological Society of America.)

normal through a large angle and so arrive in the shadow zone, as shown in Figure 2.8, rays E and F. It is not clear, even today, just what form the transition from outer to inner core takes, though there must be an overall increase in V_p; Figure 2.7b shows some suggestions. The reason for this uncertainty is partly because the rays have to travel a long way before they reach the inner core and therefore are weak, and partly because any errors in the velocity profile of the upper parts of the Earth will tend to obscure the details due to the inner core boundary. The one point that all the suggested profiles have in common is that the outer core–inner core transition is not a single discontinuity, and therefore we term it a transition zone. (The reader can find further discussion in Chapter 1 of Jacobs 1975.)

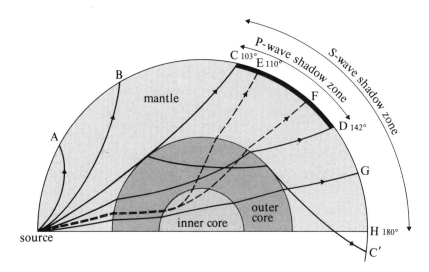

Figure 2.8 Selected P-rays
and the core shadow zone. A
to H are rays leaving the
source at progressively steeper
angles. A ray only minimally
steeper than C, which grazes
the core, is refracted into the
core and arrives at C′. The
interval C to D would receive
no rays were it not for the fact
that the inner core deflects
some into this shadow zone (a
little seismic energy gets
beyond C by diffraction, see
Fig. 8.20). (Based on a figure
in *Elementary seismology* by
C. F. Richter. © 1958 W. H.
Freeman & Co.)

It was long thought likely that the increase in V_p is due to the inner core being
solid, for, if the rigidity modulus, μ, is not zero there, it would account for an
increase in V_p relative to the outer core, where μ is zero (Eqs (2.2)). This would also
imply that S-waves can travel through the inner core. Calculations of the periods of
oscillation of the whole Earth (Section 2.5) supported the inner core being solid, but
the most direct proof is to detect S-waves; however, this was not achieved until 1972
(Julian *et al.* 1972). The existence of a solid inner core is important, for instance, to
our understanding of convection in the core, which seems to be necessary to account
for the existence of the Earth's magnetic field (Ch. 6).

Turning to the mantle, the velocity shows an overall increase with depth, upon
which are superimposed a number of steps in the top 1050 km. In Chapters 3 and 7
we shall identify these with phase changes, due to the great pressure converting
minerals to denser forms with larger moduli. There is also evidence of a transition
zone at the base of the mantle, for the seismic velocities there increase less rapidly
than in the regions above.

Another remarkable feature in the mantle is the decrease in velocity between
about 60 and 250 km depth, called the **low-velocity zone**. It causes a shadow zone (a
shadow zone is a consequence of a downward decrease in velocity) and other unusual
properties, among which is that waves passing through it are attenuated, i.e.
they are partly absorbed. The low-velocity zone is often identified with the
asthenosphere, a semi-plastic layer that permits vertical movements or isostatic
adjustments of land masses to occur (Section 8.2). However, as the seismic and
plastic properties of a rock are not closely related, the two regions should not be
thought of as being synonymous. This will be considered in more detail in Section
8.5.

Finally, so close to the surface that it hardly shows on Figure 2.7, is the Moho (a
convenient abbreviation of the Mohorovičić discontinuity, named after the
Yugoslavian seismologist who discovered it in 1909). Although there is usually an
abrupt increase of seismic velocity on descending through the Moho, there is
considerable diversity in the velocities immediately above and below the discontinuity. The Moho occurs nearly everywhere but its depth varies, being typically
35 km beneath the continents (with a range 20–90 km), and 5 to 10 km beneath the

ocean floor. It defines the boundary between crust and mantle, and the nature of the change of rock across it will be discussed in Section 7.2.

These are only the more obvious pieces of information evident from the velocity–depth profiles (Fig. 2.7). The profiles should also be able to tell us something about the density and elastic properties of the regions between the discontinuities. However, because this cannot be achieved without additional data, we turn our attention to other types of seismic information.

2.4   Surface waves

Figures 2.1c and d illustrate the two types of seismic surface wave. **Rayleigh waves** have many similarities with waves on water, for passage of the wave causes a point on the surface of the Earth to move in an ellipse in a vertical plane, as does a cork on water. Again like water waves, the amplitude of particle motion decreases with depth below the surface. This is an advantage to submarines for, if they submerge to 100 m or so, they are hardly disturbed by the surface waves produced by a storm. The amplitudes of waves of different wavelengths fall off with depth in the same way, provided the depth is measured in terms of wavelength. Thus, we have the important fact that the disturbance due to long wavelengths makes itself felt at greater depth than that due to short waves.

The velocity of Rayleigh waves depends chiefly on the rigidity modulus and on the density of rock, and usually may be taken to be about 0·92 of the velocity of S-waves. Because their disturbance is not confined to the actual surface, their velocity is not determined entirely by the properties of the surface rocks, but also by those at depth, though in a progressively decreasing way.

This dependence of velocity upon the properties at depth will be larger for longer waves, since their disturbance extends to greater depths. Consequently, in a region where V_S varies with depth – as is usually the case – the longer wavelengths will have a velocity different from that of shorter waves, and this variation of velocity with wavelength can be used to determine how V_S varies with depth.

The velocity determined by measuring the time it takes surface waves of a given wavelength to travel between two places on the surface of the Earth must be an average velocity. This may be an advantage, compared to body wave results, if we wish to ignore small-scale variations such as those due to the great variability of crustal rocks. Another advantage is that dispersion yields data about areas, such as the oceans, where there are very few recording stations. Figure 2.9a shows results for some different types of area, and Figure 2.9b shows the variation of V_S with depth that has been deduced from the surface wave dispersion. It should be appreciated that, although these results demonstrate regional differences, the departures from spherical symmetry are quite small. The subject of lateral variation will be considered further in Section 8.9.

So far, we have considered only Rayleigh surface waves. The second type of surface wave is the **Love wave**, which has motion entirely horizontally (Fig. 2.1d). The properties of Love waves are rather more complex than those of Rayleigh waves, but, as regards the information we shall derive from them, they can be considered as complementary to Rayleigh waves.

19

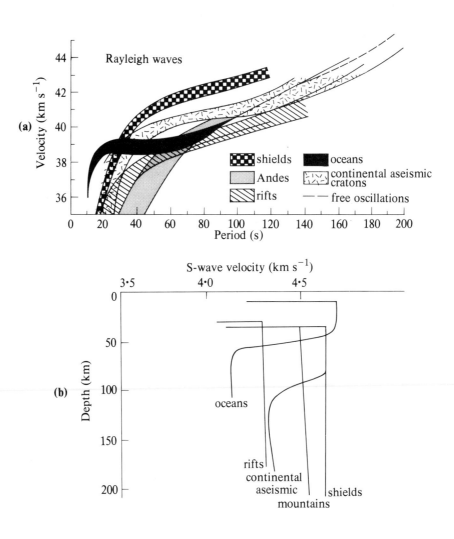

Figure 2.9 Dispersion of surface waves. (a) Dispersion of Rayleigh waves in regions with different structures. (The period of a wave is a measure of its wavelength, since it equals the velocity divided by the wavelength.) Free oscillation data derives from waves whose lengths approach the size of the Earth. (b) S-wave velocity profiles consistent with (a). Love waves give similar results. (Based on a figure by Knopoff, 1972, by permission of Elsevier Scientific Publishing Co.)

2.5 Free oscillations of the Earth

Any elastic body can be set into characteristic oscillations by a suitable disturbance, as for example a bell, a violin string or the column of air inside an organ pipe. The Earth, too, is elastic and can be set into natural oscillations by a large earthquake, and may continue to vibrate or ring for many hours or even days.

The calculation of these frequencies is a formidable mathematical problem, and so this section attempts to give a qualitative understanding of what is involved without setting up and solving the appropriate equations. The boxed pages that follow describe the principles; those who prefer to omit it will find a summary immediately after, at the end of the section.

When any elastic body has waves of some sort propagated through it, or over its surface, then, in general, these waves will be reflected at the ends or boundaries of the body. This can be verified by watching water waves hitting a harbour wall or the end of the bath. For simplicity, if we consider parallel waves moving across a straight-sided canal, then, after reflection at the opposite sides in turn, they will be moving in the original direction (Fig. 2.10a). They may then coincide with new waves being generated, in which case the wave amplitude will be increased, or they may be 'out of phase' and so partly cancel. The conditions for reinforcement is that an exact number of wavelengths can be fitted into twice the width of the canal (Fig. 2.10b).

When this condition is met, the waves no longer appear to move across the canal but remain fixed, as can be seen by adding together the outward and reflected waves at successive instants (Fig. 2.10c). For this reason, they are called **stationary** or **standing waves**. This is the same result as we would have got if we had solved the equations for oscillations in the canal without considering travelling waves; the two approaches are equivalent.

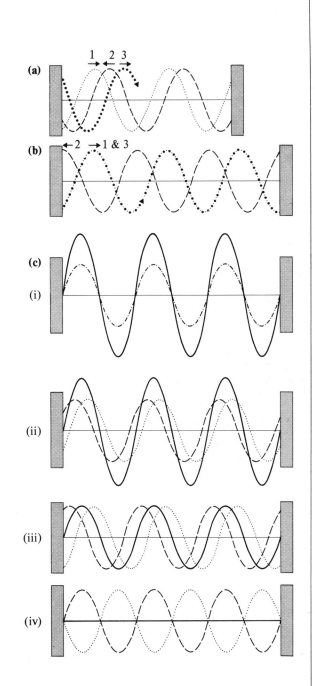

Figure 2.10 Standing waves. (a) A water wave travels outwards (1) and then is reflected back (2) by the right-hand wall. In general, after reflection at both walls, the wave (3) does not coincide exactly with the original wave, and so they partly cancel. This cancellation is complete after many reflections. (b) If an exact number of wavelengths fits into twice the distance between the walls, the doubly reflected wave coincides with the original one and enhances it. (c) These four diagrams show the addition of any wave and its reflection at different times, under condition (b). The amplitude of the combined waves (solid line) varies from nothing to twice that of either component wave, but the positions of its crests and troughs do not move, hence the term 'standing wave'.

21

The same reinforcement to form standing waves occurs in a two-dimensional body such as a sheet of metal that is struck (e.g. a gong), except that the waves will reflect several times from different edges before regaining their original direction. Similarly, we can have standing waves in a three-dimensional body, which is why a bathroom singer finds certain notes enhanced. In all cases, we find that, when we strike a body, certain wavelengths persist because they reinforce themselves, whereas wavelengths between rapidly die away. It should be appreciated that many separate wavelengths can occur simultaneously, provided they each meet the conditions for reinforcement. For example, wavelengths equal to twice the width of the canal, the width of the canal, one-half, one-third, one-quarter the width, . . . all will reinforce and form standing waves. Normally, many will be found, though the longest ones usually will be predominant.

In the case of the Earth, surface waves, rather than body waves, are important in producing standing waves. The ones that most interest us are the very long ones, comparable to the radius of the Earth, because their amplitudes fall off most slowly with depth and so are affected by the elastic moduli and density at great depths, in some cases even within the inner core. These wavelengths are far longer than those discussed in the previous section, which were affected by the properties of only the top few tens of kilometres. The distinction between surface waves and free oscillations is not only one of wavelength but that, in the latter case, we are dealing with standing waves. In general, it is more convenient to think in terms of modes of oscillation rather than wavelengths, and what we measure are the periods of oscillation, often called **eigenperiods**. The oscillations are described as free because they are the ones that are natural to a body and that persist if it is free of external forces which could impose other motions, i.e. forced oscillations.

The pattern of standing waves causes the surface of the Earth to be divided into adjacent areas which at any instant are moving in opposite senses, separated by lines (**nodes**) where there is no movement at all (Fig. 2.11). Particle motion may be predominantly radial, i.e. up and down, or sideways, and are called, respectively, spheroidal and toroidal oscillations. Figure 2.12 shows some of the slowest types of oscillation, with few nodal lines, and corresponding to very long surface waves. (The $_0S_1$ oscillation does not occur as it would involve the *whole* Earth moving back and forth; for this to happen in an isolated body would be equivalent to picking oneself up by one's bootlaces!)

Table 2.1 gives the periods of the slowest oscillations. In general, the simpler the oscillation the longer the period ($_0S_2$ is an exception, for reasons that will not be discussed here) and hence the longer its equivalent wavelength, and therefore the more its period is affected by the physical properties at depth. It is because the different modes of oscillation penetrate' to different depths that they are a useful tool for investigating the Earth's interior.

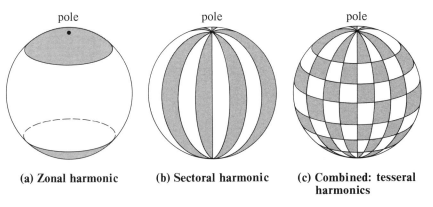

Figure 2.11 Surface movements of some oscillations of the Earth. Light and dark areas have displacements which are in opposite senses at any instant, and are separated by nodal lines where there is never any movement (nodal lines are where nodal planes within the Earth intersect its surface). Movements corresponding to the left-hand diagram are shown in Figure 2.12 ($_0S_2$ mode).

(a) Zonal harmonic   **(b) Sectoral harmonic**   **(c) Combined: tesseral harmonics**

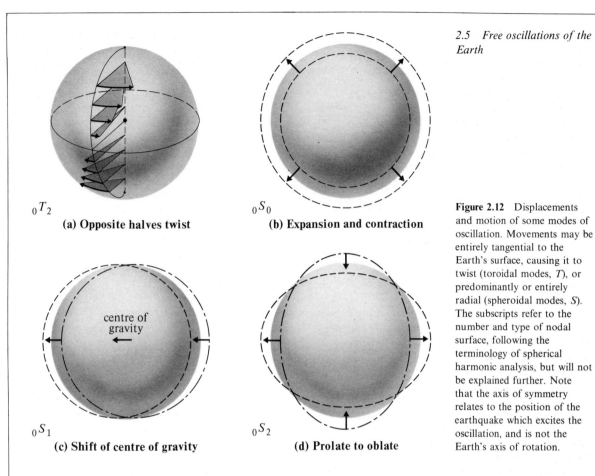

$_0T_2$

(a) Opposite halves twist

$_0S_0$

(b) Expansion and contraction

centre of gravity

$_0S_1$

(c) Shift of centre of gravity

$_0S_2$

(d) Prolate to oblate

Figure 2.12 Displacements and motion of some modes of oscillation. Movements may be entirely tangential to the Earth's surface, causing it to twist (toroidal modes, *T*), or predominantly or entirely radial (spheroidal modes, *S*). The subscripts refer to the number and type of nodal surface, following the terminology of spherical harmonic analysis, but will not be explained further. Note that the axis of symmetry relates to the position of the earthquake which excites the oscillation, and is not the Earth's axis of rotation.

Table 2.1   Periods of free oscillation (eigenperiods) of the Earth. This is a small selection of the modes observed. Errors are less than 0·1%. (From a compilation by Anderson & Hart 1976.)

Spherical modes		Toroidal modes	
Mode	*Period* (min)	*Mode*	*Period* (min)
$_0S_0$	20·46		
$_0S_2$	53·83	$_0T_2$	43·94
$_0S_3$	35·56	$_0T_3$	28·37
$_0S_4$	25·76	$_0T_4$	21·72
$_0S_{10}$	9·67	$_0T_{10}$	10·31
$_0S_{20}$	5·792	$_0T_{20}$	5·993
$_0S_{40}$	3·538	$_0T_{40}$	3·333

The meaning of the symbols $_0S_0$, etc., is outlined in Figs 2.11 and 2.12.

23

In summary, the periods of oscillation of the Earth are determined by the variation of density and elastic moduli within the Earth. These are the same parameters that determine the body wave velocities, but oscillations of the Earth are a completely independent way of investigating them, and so complement body wave studies.

Summary

1   Seismic body waves are studied by the time it takes them to travel through the Earth and re-emerge at different distances from their source, sometimes after one or more reflections or refractions at interfaces within the Earth. This information shows that the Earth to a very good approximation is spherically symmetrical (after correction for the equatorial bulge).

2   Body wave travel-time curves for P- and S-waves can be constructed and inverted to deduce the seismic velocities V_p and V_s at all depths. Discontinuities and irregularities in these velocity–depth profiles show the existence of crust, mantle and core, and other features (Fig. 2.7).

3   Surface waves are a useful way of investigating near-surface differences averaged over different areas, such as between ocean, shield and mountain.

4   Free oscillations of the whole Earth – or standing surface waves where wavelengths are comparable to the radius of the Earth – depend upon the same physical parameters (compressibility and rigidity moduli, density) as do body wave velocities, but are entirely independent. They are, therefore, a complementary source of information about the variation of these quantities within the Earth.

5   Although neither a knowledge of oscillation periods nor of body wave velocities by themselves can be used to deduce density within the Earth, they are a vital constraint which, with other information, set close limits on the possible range of density at any depth. This will be the subject of the next chapter.

Further reading

General books:
Garland (1971); Stacey (1977): seismology with a mathematical treatment.
Jacobs *et al.* (1974): global seismology (but not including oscillations of the Earth).

3 The density within the Earth

3.1 Introduction

Seismology has told us much about the layered structure of the Earth, but little about the physical or chemical properties of the layers. Of course, the body wave seismic velocities depend upon the density and two elastic moduli, but because we have only two equations to solve for these three unknowns (Eqs (2.2)) it is not possible to deduce these quantities from the P- and S-wave velocities alone. The periods of free oscillation provide extra equations which, in principle, allow us to solve for ρ, K and μ but, in practice, their use is limited, for reasons to be discussed in Section 3.6.

To help solve this problem, we measure two other quantities that depend upon the density within the Earth, namely, the total mass and the moment of inertia.

3.2 The mass of the Earth

We determine the Earth's mass, M_E, from the gravitational field that it produces. Newton's law of gravity states that the attractive force F between two point masses is

$$F = G\frac{m_1 m_2}{r^2} \tag{3.1}$$

where m_1 and m_2 are the two masses at a distance r apart, and G is the universal gravitational constant. If this equation is applied to all the particles in a spherical shell, it can be shown that the force the shell exerts *outside* itself is the same as if all its mass were concentrated at its centre. Thus, a body that consists of concentric layers – as does the Earth, to a good approximation – has an external attraction which is symmetrical and falls off as the inverse square of the distance from its centre. It is convenient to measure this attraction at any point by the force it would exert upon a unit mass at that point. If we put m_1 equal to unity and m_2 equal to M_E, the mass of the Earth, we have the acceleration due to gravity (or force on unit mass) g_r as

$$g_r = G\frac{M_E}{r^2} \tag{3.2}$$

It is an acceleration because with only one mass in the equation it has the dimensions of acceleration, and it is the acceleration with which a body would fall. When referring to this acceleration at the surface of the Earth, the suffix r will be omitted.

One method of determining M_E is to measure g at the surface of the Earth, by timing the free fall of a mass in a vacuum. The quantity r in Equation (3.2) is the radius of the Earth and is determined by surveying. The constant G can be determined in the laboratory by measuring the force of attraction between two masses. This was first done by Cavendish in 1798, who went on to calculate the mass of the Earth.

A second method of determining M_E is to measure the period of rotation of a satellite about the Earth. If, for simplicity, we suppose it is in a circular orbit (Fig. 3.1), the centrifugal and gravitational forces are always equal and opposite

$$G\frac{M_E m}{r^2} = \frac{mV^2}{r} \qquad (3.3)$$

where m is the mass of the satellite and V is its speed. The mass m 'cancels', and since the time T that the satellite takes to make one revolution is

$$T = \frac{2\pi r}{V} \qquad (3.4)$$

the velocity V in Equation (3.3) can be replaced and the equation rearranged to give

$$M_E = \frac{r^3}{G}\left(\frac{2\pi}{T}\right)^2 \qquad (3.5)$$

The period T is easily observed, and r can be measured by radar or by laser. (Equation (3.5) still holds for the more general case of an elliptic orbit.)

The value found for M_E is 5.98×10^{24} kg. Since we know the Earth's radius, we can calculate its approximate volume, and we obtain a mean density of 5520 kg m^{-3}. As the densities of most surface rocks lie in the range 2500 to 3000 kg m^{-3}, it is obvious that the inner parts of the Earth must be much denser than the outer parts. But because any arrangement of concentric shells of different densities, *having the same total mass*, produces the same external gravitational field it is not possible to deduce from its gravitational field how the density varies radially within a spherical body, and therefore we turn to other constraints. (To the extent that the Earth is not perfectly symmetrical, it is possible to learn something of its interior inhomogeneities from the variation of gravity over the surface of the Earth. On a small scale, this can be used to interpret geological structures, and on a global scale will be discussed in Section 8.9.)

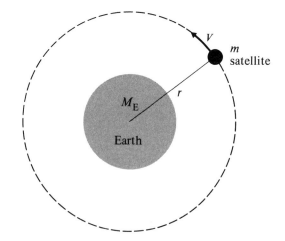

Figure 3.1 Circular satellite orbit. The mass of the Earth, M_E, can be calculated from the distance and orbital period of the satellite. See text for details.

26

The Earth's **moment of inertia** (see Note 2 for an explanation of this and related terms) is another quantity that depends upon the mass in its interior but unlike M_E it is sensitive to its distribution. The way in which it is deduced is described in the following boxed pages; those who wish to omit this will find the result summarised at the end of Section 3.3.2, following the box.

3.3.1   *The shape of the Earth*

In the laboratory, the moment of inertia of a body can be determined by measuring the angular acceleration produced by a known torque or couple: the greater the moment of inertia, the smaller the angular acceleration. In the case of the Earth, we have to rely on natural torques, and these are produced by external bodies, chiefly the Moon and Sun. Since they exert a torque only because the Earth is not quite spherical, it is necessary to discuss the shape of the Earth.

The Earth has a slight equatorial bulge because of its rotation. To predict this shape accurately by calculation, we would need to know the distribution of density within its interior. But this density variation is what we wish to find out, and so, instead, we adopt a practical definition for the shape of the Earth. This is the **geoid** which is defined as the mean sea level

surface; in continental areas, one imagines sea level canals to define the geoid. The shape of the geoid depends upon variations of the Earth's gravitational attraction. To see what this statement implies, consider a body made of concentric layers of liquid of different densities. As long as the body is not rotating, these layers will be spherical shells and the force of gravity will be directed towards the centre of the body, and perpendicular to its surface. Thus, no work would be done against gravity in moving a boat around on the surface. We call such a surface a gravitational **equipotential**, and in this case it would be spherical. Next, let the body rotate. At any place (other than the poles) there will now be a centrifugal force, C, due to the rotation, directed away from the rotation axis (Fig. 3.2). If the body were still a sphere, this extra force would cause the boat to move towards

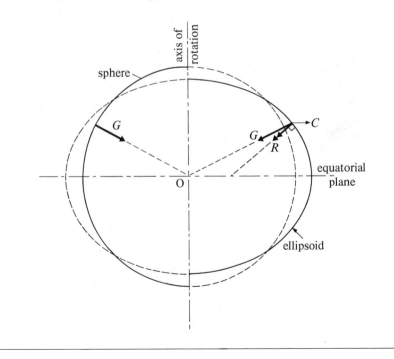

Figure 3.2 Shape of a rotating ball of liquid. *Left*: if the ball were not rotating, it would be a sphere with the vertical, G, (defined by a plumb-line) passing through the centre of the ball. *Right*: rotation adds centrifugal force, C, and the ball deforms until the combined gravitational and centrifugal forces are perpendicular to the surface, R. This new vertical does not, in general, pass through the centre.

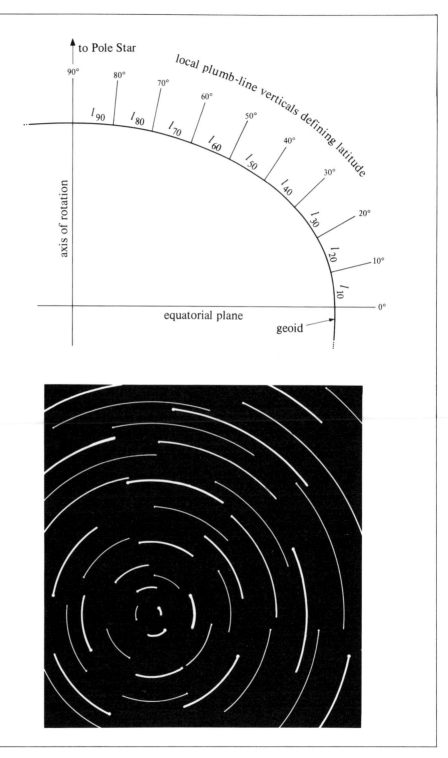

Figure 3.3 Determination of shape of the geoid. As the Earth is slightly flattened, the lines of latitude – deduced from the angle between the local vertical and the Pole Star or other convenient star – are not an equal distance apart, i.e. $l_{90} > l_{80} > l_{70}$ etc. Note that verticals in general do not point to the centre of the Earth. The cross section is very close to an ellipse.

Figure 3.4 Time exposure of the night sky. In the course of several hours, each star traces an arc about a point. This point is the celestial pole and lies on the Earth's axis of rotation, when extended. It is, therefore, exactly above the North Pole. The Pole Star is close to the celestial pole and so traces out only a very small arc.

28

the equator. The liquid would move too and must continue to do so until the body takes up a new shape such that everywhere the combined force, R, due to gravity, G, plus centifugal force, C, is exactly perpendicular to the surface. Note that the vertical, i.e. the direction of the combined gravitational and centrifugal forces, defined by a plumb-line, will no longer point towards the centre (except at the poles and equator). This reasoning applies to the Earth, for the strength of its material is negligible compared to the other forces involved (see Section 8.3) and so it can be treated as a liquid for this purpose.

The geoid can be determined by surveying, as illustrated in Figure 3.3. If some particular star is exactly overhead at the North Pole – as is almost true of the Pole Star – then, during the course of a night, all other stars will appear to travel round it in circles, as can be demonstrated by exposing a film for some hours (Fig. 3.4). This allows the direction of the true pole star ('the celestial pole') to be deduced at any latitude, and this can be compared with the local vertical. The difference determines the latitude; for example, if the difference were 20°, the latitude

would be 70°. Because the geoid is not quite spherical, the distances between lines of latitude, when measured in kilometres instead of degrees, differ slightly (Fig. 3.3). Nowadays, better determinations of the geoid come from observing satellites (see Section 3.3.2).

The problem of determining the shape of the Earth is not a simple geometrical one, but is intimately bound up with gravity. This is because surveying instruments rely upon verticals determined from spirit levels or plumb-lines, and this dependence upon gravity extends to satellite orbits. Photographs of the Earth taken from space are not accurate enough to determine its geometrical shape, though they do show the equatorial bulge. In any case, the geoid is a more useful definition.

The geoid is found to be very close to an ellipsoid of revolution (Fig. 3.3). The ellipticity is only 1 part in 298, so that the equatorial radius of 6378 km is 22 km longer than the polar one. The deviations of the geoid from this approximation are measured in tens of metres only, and so are negligible for present purposes, though they will be discussed in Section 8.9.

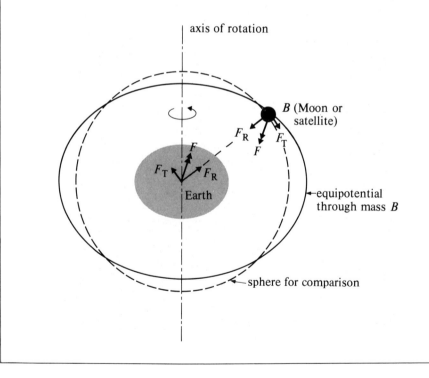

Figure 3.5 Gravitational equipotentials outside the Earth. For details see text.

29

3.3.2 *Deduction of the Earth's moment of inertia*

Because of the equatorial bulge, not only the geoid but all gravitational equipotentials outside the Earth will be slightly non-spherical (Fig. 3.5). Since the gravitational force is, by definition, perpendicular to the equipotential, the attraction between the Earth and any body B in space, such as the Moon, Sun or a satellite, will not be directed exactly towards the Earth's centre (unless they lie in its equatorial plane, or on its axis). This attraction, F, is equivalent to a large force F_C along the line of centres and a much smaller one F_T at right-angles. From Newton's third law of motion, that action and reaction are equal but opposite, it follows that the Earth must experience similar forces due to the body. The attraction F_R along the line of centres is balanced by the centrifugal force, but the small forces F_T are not balanced and together they form a torque *tending* to move the body clockwise and to rotate the Earth anticlockwise.

We can gain a different understanding of this torque, without using the concept of equipotentials, by considering a highly simplified model. In Figure 3.6, the equatorial bulge has been replaced by two equal masses m_1 and m_2 at the ends of the diameter. The attractions between these two masses and B are not quite the same because of their different distances (Eq. (3.1)). As drawn, this slight inequality of F_1 and F_2 will produce a small anticlockwise torque upon the Earth, and therefore, by reaction, a clockwise torque upon the body B. We could calculate these forces using Equation (3.1), and hence the torques, provided we knew the values of all the masses and distances involved. This would give the same result as if we first calculated the equipotentials near B and then calculated the forces that are perpendicular to the equipotentials. In practice, we use the potential method because it is not easy to replace the equatorial bulge by a number of point masses. The point to note is that the greater the moment of inertia of the Earth about its rotation axis, the greater is the torque. This is because the larger the moment of inertia, the more the mass of the Earth must be distributed away from the axis; in turn, the further away the mass is from the axis, the greater is the centrifugal force that it will experience and hence the greater is the bulge.

The torque also depends upon the mass and distance of the external body. In the case of an artificial satellite, the torque is extremely small and so has negligible effect upon the Earth. However, the effect of the Earth upon the satellite is very noticeable, perturbing its orbit, and this allows us to make deductions about how the Earth's attraction varies in space and hence about the geoid, as mentioned in the previous subsection. The largest torques upon the Earth are produced by the Moon and, to a lesser extent, the Sun, for, though the Sun is larger, its greater distance more than outweighs this.

It would seem obvious that the effect of the torque of the Moon (or Sun) upon the Earth's equatorial bulge would be to move both bodies until the Moon lay in the Earth's equatorial plane, where the torque is zero. This, indeed, would be the case if the Earth were not rotating, but because it is rotating it behaves like a spinning top, or one of those toy gyroscopes that wobble around on their stands, apparently in defiance of gravity. In Figure 3.7, the weight of a top and the reaction at its tip produce a torque that would cause the top, if *not* rotating, to fall on its side. But we know that when it is spinning it wobbles or **precesses** so that its axis traces out a cone. Similarly, because of the torque due to the Moon, the Earth's axis is not fixed in space but precesses slowly, with the axis of the cone perpendicular to the plane of the Earth's orbit.

If the above considerations are followed through quantitatively, a mathematical expression for the rate of precession can be derived (see, e.g., Garland 1971, Stacey 1977). It contains quantities such as the mass and distance of the Moon (or other body), plus two additional terms. One of these shows how the torque, and hence rate of precession, depends upon the deviation of the Earth's gravitational field from perfect symmetry, and this we know from the motions of satellites or other determination of the geoid. The other term shows that the rate of precession is inversely proportional to the Earth's moment of inertia about its rotation axis. Thus, the Earth's moment of inertia can be calculated from its measured rate of precession.

This rate is small because of the smallness of the bulge and the great distance of the Moon, and it takes about 26 000 years to complete one revolution. This can be detected because, though the Earth's axis at present points almost at the Pole Star (Fig. 3.4), its direction is slowly changing.

30

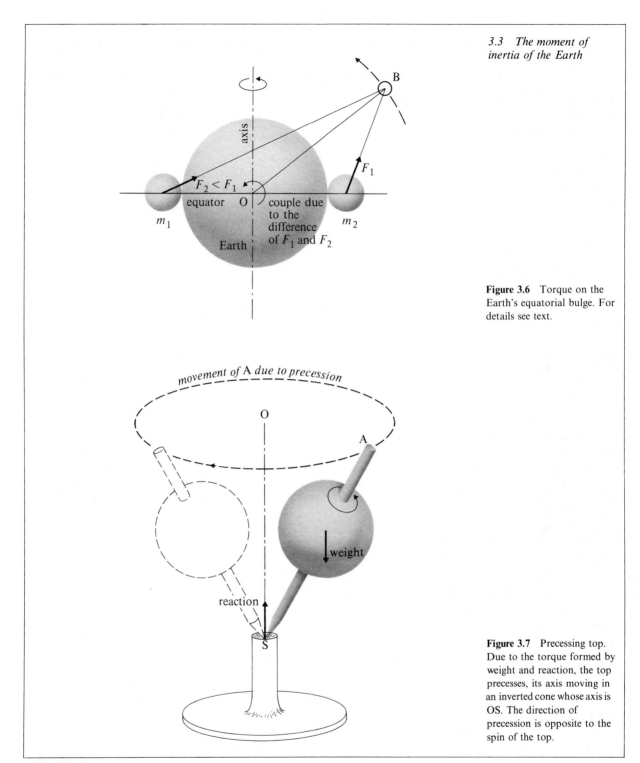

Figure 3.6 Torque on the Earth's equatorial bulge. For details see text.

Figure 3.7 Precessing top. Due to the torque formed by weight and reaction, the top precesses, its axis moving in an inverted cone whose axis is OS. The direction of precession is opposite to the spin of the top.

The Earth's moment of inertia is $8 \cdot 07 \times 10^{37}$ kg m^2. This is only 83% of the value it would have if it were of uniform density, and again tells us that the mass must be concentrated towards the centre. However, it does more than just confirm the conclusion we made from the mass of the Earth, because the moment of inertia

31

depends upon the radial variation of density, whereas the total mass does not. Expressed mathematically, we can write for the mass M_E and the moment of inertia C, the expressions

$$M_E = \int_0^{R_E} \rho(r)\, 4\pi r^2\, \mathrm{d}r \tag{3.6}$$

$$C = \int_0^{R_E} \tfrac{2}{3} r^2 \rho(r)\, 4\pi r^2\, \mathrm{d}r \tag{3.7}$$

Thus, though neither quantity can be used to deduce in detail how density varies radially, each is able to rule out many density profiles; and together they rule out many more than separately. They are, therefore, useful constraints on possible density profiles.

3.4 The simple self-compression model: the Adams–Williamson equation

The equations for the velocities of P- and S-waves, namely

$$V_p = \sqrt{\frac{K + \tfrac{4}{3}\mu}{\rho}} \qquad V_s = \sqrt{\frac{\mu}{\rho}} \tag{3.8}$$

cannot be solved for K, μ and ρ because there are three unknowns but only two equations. A third independent relationship between the variables is needed but, unfortunately, no rigorous one exists. Early attempts to deduce the density tried to get round this obstacle either by assuming a relationship or by trying to discover one empirically. One approach was that of Adams and Williamson in 1923 and, although of limited applicability, it is instructive because of the insights it gives.

They assumed that density increases with depth only because of compression due to the weight of material above and not, for instance, because of a change of composition.

The quantity that defines the change of density due to pressure is the compressibility modulus, K (Section 2.1), given by

$$K = \frac{\text{compressional stress}}{\text{volumetric strain}} = \frac{\text{increase in pressure}}{\substack{\text{resulting proportional} \\ \text{decrease in volume}}} = \frac{\mathrm{d}P}{\mathrm{d}v/v} \tag{3.9}$$

Since volume and density are inversely related, we can write

$$\frac{\mathrm{d}\rho}{\rho} = \frac{-\mathrm{d}v}{v} \tag{3.10}$$

(e.g. decreasing the volume of a given mass of material by 1% increases the density by 1%). Therefore, compressibility modulus K is

$$K = -\rho \frac{\mathrm{d}P}{\mathrm{d}\rho} \tag{3.11}$$

The extra pressure on descending through a spherical shell of thickness $\mathrm{d}r$ (Fig. 3.8), due to the extra weight above, is

$$\mathrm{d}P = -\rho_r g_r \mathrm{d}r \tag{3.12}$$

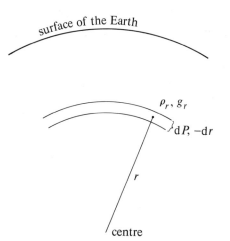

Figure 3.8 Pressure and
density within the Earth.
Increasing depth by δr
increases pressure by δP due
to the extra weight above. See
text for further details.

(The negative sign is because P increases as r decreases.) Of course, ρ_r and g_r are not absolutely uniform through the thickness of the shell, but they change much less rapidly than P. By making the shell infinitely thin (the calculus limit), the error in assuming ρ_r and g_r to be uniform can be made negligibly small. The density increase, produced by the increase in pressure, is found by substituting Equation (3.12) into Equation (3.11). The result is rearranged to give

$$\left(\frac{K}{\rho}\right)_r = g_r \rho_r \frac{dr}{d\rho} \tag{3.13}$$

The seismic velocity information in Equations (3.8) has not yet been used. These two equations can be combined to give

$$V_p^2 - \tfrac{4}{3} V_s^2 = K/\rho \tag{3.14}$$

(μ has been eliminated). Hence Equation (3.13) can be rewritten as

$$(V_p^2 - \tfrac{4}{3} V_s^2)_r = g_r \rho_r \frac{dr}{d\rho} \tag{3.15}$$

The suffix r denotes that the values of V_p, V_s, etc., are those occurring at the radius r. It should be appreciated that the elastic moduli, as well as the density, increase with depth. This must be so because a density increase alone would produce a decrease of V_p and V_s with depth (see Eqs (3.8)), whereas we know they increase. It may seem a contradiction that the elastic moduli, usually regarded as constants, can vary with depth. In fact, they are only approximately constant, but the variation does not show up over the usual range of stresses used in laboratories or encountered in engineering.

Equation (3.15) still contains g_r which is unknown and so must be eliminated. This can be done because g_r depends upon the radial density variation. We have already seen that outside a shell the gravitational attraction is the same as if its mass were concentrated at its centre. It can also be shown from Equation (3.1) that anywhere inside a shell there is no attraction. Thus, the acceleration due to gravity at a radius r is the same as if all the mass inside r were concentrated at the centre, the mass outside being ignored, i.e.

$$g_r = \frac{G}{r^2} \times (\text{sum of masses of all shells inside } r)$$

or

$$g_r = \frac{G}{r^2} \int_{r_1=0}^{r_1=r} 4\pi\rho_1 r_1^2 \, dr_1 \tag{3.16}$$

where $4\pi r_1^2 dr_1$ is the volume of each shell at distances r_1 from the Earth's core, and ρ_1 is the density at the same distance.

Finally, substituting Equation (3.16) into (3.15) and rearranging gives

$$\frac{d\rho}{dr} = \frac{G}{r^2} \frac{\rho_r}{(V_p^2 - \frac{4}{3}V_s^2)_r} \int_{r_1=0}^{r_1=r} 4\pi\rho_1 r_1^2 \, dr_1 \tag{3.17}$$

This untidy expression is the **Adams–Williamson equation**. How does it help us? First, we see that since V_p and V_s are known at all values of r (Ch. 2), the only unknown is how ρ varies with r. Secondly, at the surface of the Earth, we also know ρ, r and the integral, which is just the mass of the Earth. Therefore, we can evaluate the right-hand side of the equation at the surface of the Earth. Since the left-hand side gives the rate at which density increases with depth, we can calculate the density at a small depth below the surface, i.e. at the base of a thin surface shell. The right-hand side can now be re-evaluated at the new, slightly smaller, value of r since the integral will equal the mass of the Earth less the mass of the shell whose density we know. And so on, for progressively greater depths.

In practice, the integration is begun at the top of the mantle, because it is known that the crust is a layer of variable thickness and density (its mass is allowed for when drawing conclusions). A density of about 3200 kg m^{-3} is chosen, based upon samples derived from the mantle, and then the density is deduced at all depths down to the base of the mantle, beyond which it would be absurd to go because clearly a major change occurs there that cannot be due to a simple compression. How, then, can the core density be deduced? This is done by guessing a density for the top of the core and then using the Adams–Williamson equation to deduce the density down the centre of the Earth. This distribution must be such that the total mass, integrated over crust, mantle and core, is equal to the known mass of the Earth, and the density for the top of the core is adjusted until this is so. The result is shown in Figure 3.9.

To check whether this density distribution is correct, it is used to calculate the moment of inertia of the Earth and compared with the known value: they are found to differ significantly. Further, it can be shown that the discrepancy cannot be due to the core, as follows. The density distribution of the mantle (plus crust) is used to calculate the mass and moment of inertia of the mantle and hence of the core, by subtraction from the known values for the whole Earth. It turns out that the resulting ratio of the moment of inertia of the core to its mass is 1.4 times that of a uniform sphere, which would imply that the mass of the core is concentrated towards its surface. As it is highly implausible that the density of the core markedly *decreases* inwards, the only conclusion is that there must be more mass in the mantle than the self-compression model predicts.

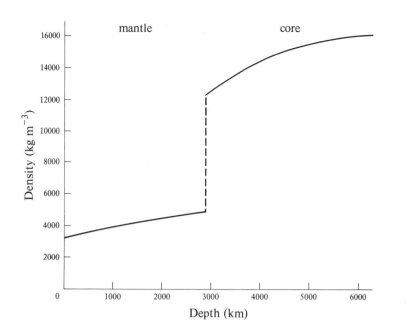

Figure 3.9 Density–depth profile in the Earth according to the self-compression model. It is assumed that the Adams–Williamson equation holds within both mantle and core, but not across the core–mantle boundary. (Curve modified from Bullard 1957.)

3.5   Probable defects of the simple self-compression model

As something is wrong with the model, the assumptions upon which it is based must be examined carefully.

One assumption is that the pressure at a given depth is equal to the weight of material above. Clearly, this is not true in a cave, because the strength of the rock supports the weight above, but at a depth of only a few kilometres the weight becomes too great for this to be possible. In fact, the strength of the Earth with respect to *long-continued* forces is negligible on a global scale, and the assumption that it behaves as a liquid, implicit in Equation (3.11), is closely correct on this time scale (Section 8.3). Corrections for departures from spherical symmetry due to inhomogeneities in the densities of rocks and the equatorial bulge also are minor.

A more important factor is temperature. Temperature does not appear explicitly in the Adams–Williamson equation, but if the temperature rises with depth, it will act in the opposite sense to pressure, tending to expand the material. It might, therefore, be thought that the Adams–Williamson equation implies a constant temperature, but this is not so because of the compressibility modulus used. There are two major definitions of this modulus, depending upon what happens to the heat that is produced when a material is compressed. In the case of the isothermal modulus, this heat is removed so that the temperature remains constant; in the adiabatic modulus the heat remains in the material, and as this tends to expand the material, a greater pressure is required to produce a given compression, i.e. this compressibility modulus is greater. The compressibility modulus used in the Adams–Williamson equation derives, via Equation (3.14), from the equation for V_p (Eq. (3.8)). When a P-wave passes through a material, a compression lasts so short a time that the heat produced does not have time to exchange with the surrounding material before the ensuing rarefaction cools it. Therefore, the appropriate modulus **35** is the adiabatic one, and so the Adams–Williamson equation implies that inside the Earth there is an **adiabatic temperature gradient**, or **adiabat** for short.

The adiabat is such an important concept that it is necessary to explain its meaning at some length. Consider a compressible liquid in which both temperature and pressure increase with depth. If the two equal small masses δm_1 and δm_2 in Figure 3.10 were to be exchanged in some magical way *without change of temperature or pressure*, δm_2 would be at both a higher temperature and a higher pressure than its new surroundings (and vice verse for δm_1). If, then, the pressure were allowed to equilibrate instantaneously with the new surroundings, δm_2 would expand, and, in so doing, it would cool. If, after pressure equilibration, the temperature were then equal to that of the surroundings, the temperature gradient would be adiabatic. If δm_2 were hotter than its new surroundings, the gradient would be superadiabatic, and if cooler it would be subadiabatic. In the superadiabatic case, the excess temperature after pressure equilibration would mean that the material would continue to expand, reducing the density below that of its surroundings, so giving it an upward buoyancy. On a large scale, a superadiabatic temperature gradient in a liquid tends to lead to convection, because if any small disturbance in the liquid displaces material upwards, the resulting buoyancy will tend to cause it to rise further (see also Section 8.6.1).

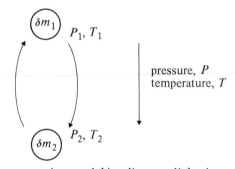

Figure 3.10 . Adiabatic temperature gradient. Two equal masses δm_1, δm_2, initially at the locations shown, are instantaneously interchanged in a hypothetical experiment. For further details see text.

Because the self-compression model implies an adiabatic temperature gradient, it follows that, if the temperature gradient in the Earth were superadiabatic, the density would rise less rapidly than predicted by the Adams–Williamson equation, because the extra temperature would cause the material to expand. Conversely, to account for the mass of the mantle exceeding that predicted by the Adams–Williamson equation would require a subadiabatic temperature gradient. As the gradient is almost certainly superadiabatic (Section 8.6.2) it must be concluded that non-adiabatic temperature gradients cannot be the major cause of the breakdown of the self-compression model.

In any case, it can be calculated that the effect of any plausible departure from adiabatic conditions would have only a small effect upon the density profile. One possible exception is the base of the mantle when the seismic velocities increase less rapidly than in the regions above. This could be due to a large temperature gradient produced by heat supplied by the core (Section 8.6.2), but alternatively might be due to a compositional change resulting in increased density. The lithosphere, the rigid uppermost part of the Earth (Section 8.2), has a large superadiabatic temperature gradient but its effect upon density is less important than that of compositional changes (Chs. 7–9).

This leaves only two other factors: compositional changes or changes of state. Since materials of different composition can have different densities at the same pressure, it is evident that the Adams–Williamson equation cannot take into account a change of material within the Earth, either abrupt or gradational. Changes of state include the change from liquid to solid, which has been inferred to occur at the core–mantle boundary, but also include **phase changes**, in which a solid material changes its crystalline structure by a repositioning of its atoms. An example is the change from graphite, density 2000 kg m⁻³ to diamond, density 3500 kg m⁻³, at high pressures. The increase of pressure with depth in the Earth will favour more compact forms and hence higher densities. Since crystals exist only as regular arrays of atoms, the transition from one form to another must be abrupt, at least on a local scale, and this too is outside the scope of the Adams–Williamson equation, for use of the bulk modulus assumes that a smooth increase in pressure produces a smooth decrease in volume. The simple self-compression model is therefore, in its essentials, equivalent to assuming a homogeneous Earth.

Whether the departures from the self-compression model are due to compositional changes or to phase changes will be discussed in later chapters. As to where the departures occur, it is likely that the seismic discontinuities or steps of Figure 2.7 are one location, because, according to the self-compression model, the density and elastic moduli should change smoothly with depth. This does not rule out that gradational compositional changes may also occur at any depth.

Though the Adams–Williamson equation is not rigorously applicable, there are regions where it applies fairly well. The outer core should give the best approximation because, being liquid and probably in convective motion, it should be well mixed with a closely adiabatic temperature gradient (see Ch. 6). The equation probably also holds approximately for most of the lower mantle, from 1050 to 2700 km. Where the equation does not hold, the density–depth gradient that it predicts should be a minimum, because both compositional and phase changes will lead to a more rapid increase of density with depth. The only likely exceptions to this are regions of very high temperature gradient, which are most likely to occur at the top and base of the mantle (Section 8.6.2), leading to low-density gradients.

3.6 More sophisticated models and the Monte Carlo inversion method

Following the realisation that the simple self-compression model has only limited applicability, other attempts were made to find a third relationship linking density and the elastic moduli, or, alternatively, the seismic velocities. From experiment, Birch (1961) proposed the relation

$$V_\mathrm{p} = a\bar{m} + b\rho \qquad (3.18)$$

where a and b are empirically determined constants and $\bar{m}$ is the **mean atomic weight** of a mineral, i.e. the sum of the atomic weights of all the atoms in a mineral, divided by the number of atoms. The physical reasons underlying this law are not fully understood, but there are doubts whether it holds at the high pressures of the deep mantle. For further details, see Note 3 and Section 7.5.

Another approach used the assumption, on the basis of the limited information available, that the compressibility modulus changed in a simple way with depth. This led to a family of models by Bullen (1963) which steadily evolved over the years to take account of improved data, and which have been widely used. A further approach was to deduce the density from proposed petrological models of the Earth (this is the reverse of the course adopted in this book). Often, the resulting models were a patchwork of assumptions as, for instance, the model of Wang (1972). In the upper mantle, Wang based the density upon a petrological model; below this he employed Birch's law (Equation (3.18)) down to the lower mantle, where he used a different empirical relation between density and P-wave velocity, while the Adams–Williamson equation was applied to the core.

Although such approaches have merit, by trying to use the best available information for each region, they are liable to be subjective.

The situation changed with the advent of free oscillation data, because the period of an oscillation depends upon the distribution of the density and elastic moduli within the Earth, and so provides an independent relationship between these quantities. In principle, a knowledge of just the eigenperiods of the Earth could yield a density–depth profile. In practice, this is not so, because of the limitation imposed by the finite number of periods observed; ρ, K and μ vary continuously within the Earth and so need to be specified at an infinite number of depths, yet the number of accurately known eigenperiods is only a hundred or so. It follows that our knowledge of eigenperiods can only constrain possible profiles of the density and elastic moduli, not determine unique ones. Furthermore, since Rayleigh wave velocities depend only slightly upon the compressibility modulus, and Love waves not at all, a knowledge of the periods of spheroidal and toroidal modes places very little constraint upon the variation of the compressibility modulus or V_p with depth.

Because of these limitations upon the use of free oscillation data, it is usual to combine them with body wave data and the mass and moment of inertia of the Earth, to constrain possible density–depth profiles. One way to do this, without assuming any model, is to use the Monte Carlo inversion method, taken to its greatest development by Press (1968, 1970a, b).

In the Monte Carlo method, profiles are chosen at random and then tested against the available constraints. In this way, preconceptions are eliminated. Press used a computer to choose randomly a value of the density at each of 19 depths, then joined successive points together to specify a profile (Fig. 3.11). To reduce the computation needed, limits were put on the possible range at each depth and, to allow for more rapid variation at some depths, the 19 selected depths were not equally spaced (Fig. 3.11). This was also done for V_p and V_s. (Because of Equations (3.8), ρ, V_p and V_s are equivalent to ρ, K and μ.) The Earth was then divided up into 81 concentric shells, with the value of ρ, V_p and V_s in each shell being found from the randomly chosen profile. The next step was to use the multi-shelled model to calculate the total mass and moment of inertia of the Earth, most of the well determined eigenperiods, and the body wave travel times for a number of selected epicentral distances (to calculate a complete travel-time curve would be too extravagant of computer time). Each of these calculated quantities was compared with the observed values and if they differed by more than the observational error – usually less than 1% – the relevant profiles were rejected and new ones chosen, again randomly.

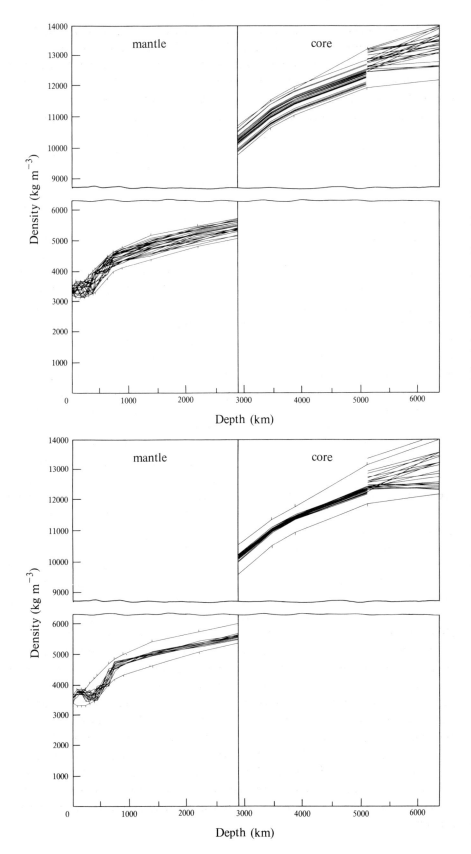

Figure 3.11 Twenty-five randomly generated density profiles. This very small selection of models tested during the application of the Monte Carlo inversion method fills the space between the limits fairly evenly, consistent with them being random. The vertical marks on the boundary lines indicate the depths at which V_p, V_s and ρ were randomly chosen. (Redrawn from Press 1970a, by permission of Elsevier Scientific Publishing Co.)

Figure 3.12 Twenty-seven successful density profiles obtained using the Monte Carlo method. At most depths, the solutions cluster fairly closely, suggesting that other possible solutions would not differ greatly. But note that they touch one or other limit at some depths, indicating that the limits are too narrow. Note also the poor constraint the solutions give for the inner core density. (Redrawn from Press 1970a, by permission of Elsevier Scientific Publishing Co.)

39

Press tested tens of millions of models, of which only 27 passed; these are shown in Figure 3.12, for density only.

How much reliance should we place on these results? The first point to appreciate is that the tens of millions of models tested are still only a small fraction of the possible models that could have been tested. Press did not state into how many intervals he divided the permitted range at each depth, but evidently it is more than 10 (Fig. 3.11). With 19 separate depths, this would yield over 10^{19} independent models for density alone. Thus, the 27 models form only a small selection of those possible, and so their significance depends upon the extent to which they are representative of the total. This can be judged only from their scatter, and Figure 3.12 shows that at most depths the scatter is considerably less than the range allows. Therefore, it seems *probable* that further successful models would not differ greatly from those already obtained, except in the low velocity zone where the successful models differ considerably.

There are more serious objections to the results. At some depths, the successful solutions crowd against the limits, even though Press claimed that these had been chosen to include all previous models. It seems likely that successful models could exist outside the limits at these points. Furthermore, if the *actual* density profile is, say, above the limit at some period then, by 'forcing' models below the limit, the mass that has been 'removed' has to go elsewhere, to satisfy the total mass and other tests. Thus, too narrow limits at one depth may result in spurious solutions at other depths, as well.

There are other objections, concerning the number of depths at which values are chosen, the methods of computation and the choice of tests and the permitted deviations. These have been criticised by Worthington *et al.* (1972) who conclude, however, that the method, in principle, can give unbiased results.

However, the method does have the inherent limitation that, to reduce computation time to an acceptable value, it is necessary to make approximations – such as specifying the density at only 19 depths – that are not forced upon other approaches.

3.7 Density versus depth: the state of play

Improvements in deducing the density-depth profile in the Earth currently rely largely upon correcting existing profiles in the light of improved data, or by removing known discrepancies. More detailed seismic velocity–depth profiles are frequently produced, resulting both from improved instrumentation and improved analysis of the recorded signals. There has also been a steady improvement in free oscillation data, both in the number of modes identified and in the precision with which their periods are determined.

Figure 3.13 shows a selection of recent density–depth profiles. The main point to be appreciated is that the differences are minor, and so some confidence can be placed in present knowledge of the density at any depth. This is not to state that the remaining uncertainties are trivial, for more details are needed in the top 1000 km of the mantle, at the base of the mantle, and also in the inner core. In the first case, we can hope to supplement our knowledge with other evidence, particularly geoche-

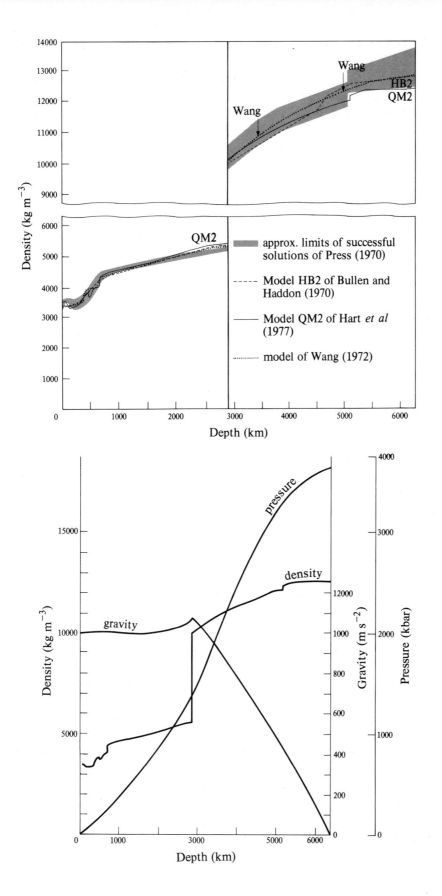

Figure 3.13 Selection of recent density–depth profiles for the Earth's interior. The assumptions of Press (1970a) and Wang (1972) are given in the text. Bullen and Haddon (1970) based their model on the Adams–Williamson equation, with modifications to accommodate free oscillation data. Hart *et al.* (1977) allow for the plastic behaviour of the Earth's mantle. (Figure redrawn from the sources quoted.)

Figure 3.14 Density, pressure and gravity within the Earth. These are recent estimates, and are adopted for this book. Pressure and gravity are from Model C2 of Anderson and Hart (1976); density is from the closely related Model QM2 of Hart *et al.* (1977), which differs from Model C2 only in allowing for the inelastic, or plastic, behaviour of the mantle. (A pressure of 1 kbar is approximately equal to 1000 atm.)

41

mical, but our knowledge of the composition of the base of the mantle and of the inner core stems chiefly from their densities.

For the remainder of this book, the density–depth profile of Hart *et al.* (1977) has been adopted; it allows for the inelastic behaviour of the mantle and satisfies the constraints of total mass, moment of inertia, free oscillation and body wave data. It is shown in Figure 3.14, together with gravity and pressure which can be deduced from it using Equations 3.12 and 3.16.

Summary

1   The seismic velocities V_p and V_s, which are known with considerable precision at most depths, depend upon the density and two elastic moduli, but it is not possible to solve for these three parameters without a third relation between them. As no rigorous relationship exists, attempts to solve for density, etc., depend upon assumptions or empirical relationships, or a combination of these.

2   Use of the simple self-compression model shows that the mantle cannot be a homogeneous material with density increasing only because of self-compression and with an adiabatic temperature gradient.

Probable regions of departure from this model are between depths of 100 and 1000 km in the mantle and at the base of the mantle, together with the mantle/core discontinuity and the outer core – inner core transition zone (Fig. 3.13).

The breakdown of the model at most of these depths is due either to compositional changes or to changes of state, which include phase changes; these will lead to a density gradient greater than that due to self-compression alone. In the lithosphere and possibly at the base of the mantle, large superadiabatic temperature gradients may produce the opposite effect, though a compositional change is an alternative explanation for the small rate of increase in seismic velocities near the base of the mantle.

3   More sophisticated models use a variety of approaches to exploit the best available data, and may be a patchwork of different assumptions for different depths. Such models are liable to have a subjective bias, and an alternative approach is the Monte Carlo inversion method. However, though this method is unbiased, in principle, the great amount of computation required limits its usefulness.

4   Despite these reservations, density is now known at most depths with a precision that is a tight constraint on composition. Depths where an improved knowledge particularly would be welcomed are in the top 1000 km of the mantle and the inner core–outer core transition zone.

Further reading

Advanced books:
   Runcorn (1967): *Dictionary of geophysics* – see under 'Earth, density distribution of' and 'Earth models based on seismology'.
   Bullen (1963): departures from the simple self-compression model (Section 3.4).
Advanced journals:
   Press (1968, 1970b): use of Monte Carlo inversion (1968); additional details and solutions (1970b).
   Hart *et al.* (1977): models for density, pressure and gravity in the Earth.

4 The formation of the Solar System and the abundances of the elements

4.1 Why it is necessary to look outside the Earth

In the two previous chapters, the density and some other parameters within the Earth were deduced with considerable precision. Next, we want to know what sort of material is there, i.e. what chemical composition it has and of what minerals it is made.

It has already been established that the Earth is layered and, in particular, that the crustal rocks are much less dense than the Earth as a whole. Therefore, though the composition of the surface rocks must be taken into account, they cannot be regarded as representative of the mantle and core, and it is necessary to look elsewhere for additional information. We are forced to ask how the Earth was formed, and, in turn, this widens the enquiry to the formation of the Solar System. (Capital letters will be used to denote the particular solar system of which the Earth is a member; similarly for other terms.) By studying members of the Solar System, it is possible to estimate its original composition and the physical and chemical processes that have led to its present state.

This chapter will be concerned chiefly with establishing the bulk composition of the Earth and Chapter 5 will concentrate on the processes that have produced a layered planet.

4.2 Introducing the Solar System

The Solar System consists of the Sun and all the bodies that orbit around it, of which the most massive are the nine planets. Table 4.1 lists many of the properties of the Sun, the planets and the Moon, and Figure 4.1 displays their relative sizes. Less tersely describable features are outlined in the following 'Who's Who in the Solar System'.

The Sun is an average star. It consists of about 70% hydrogen, 28% helium and only 2% of heavier elements, and it generates its heat and light by converting hydrogen to helium in its deep interior (Section 4.6.2).

Mercury is the planet nearest to the Sun. It is quite small and has a density only a little less than the Earth's. Its surface is heavily cratered, like the Moon's.

Venus has been called the sister of the Earth because its mass and density are only a little less, but its spin is slow and retrograde (i.e. opposite to the direction of revolution of all the planets about the Sun). Its atmosphere is much denser than ours, with a temperature at the planet's surface of about 500°C, and it consists chiefly of carbon dioxide, with some water and such unpleasant substances as sulphuric and hydrochloric acids. We cannot see the surface, but radar shows it to be cratered.

Table 4.1 Chief properties of the Sun, the planets and the Moon.

Property	Sun	Terrestrial planets					Major planets				
		Mercury	Venus	Earth	(Moon)	Mars	Jupiter	Saturn	Uranus	Neptune	Pluto
distance from Sun (mean value) (units of 10^6 km)	—	58	108	150	—	228	778	1427	2870	4497	5900
(Earth=1)	—	0·39	0·72	1	—	1·52	5·20	9·54	19·2	30·1	39·4
mass (Earth=1)	343 000	0·055	0·815	1	0·012	0·108	318	95	14·6	17·2	c.0·002
mean density (water=1)	1·4	5·4	5·2	5·5	3·3	3·9	1·3	0·7	1·2	1·7	< 1·7
radius (km)	696 000	2440	6052	6378	1738	3394	71 400	60 000	25 900	24 750	1900
year, i.e. period of revolution about Sun (Earth years)	—	0·24	0·62	1	—	1·88	11·9	29·5	84·0	164	248
spin period, i.e. rotation about axis (days)	27	59	−243*	1	27·3	1·03	0·40	0·43	−0·89*	0·53	6·4
eccentricity of orbit	—	0·206	0·007	0·017	0·055	0·093	0·043	0·056	0·047	0·009	0·25
inclination of orbit, with respect to the Earth's (deg)	—	7	3·4	0	23†	1·9	1·3	2·5	0·8	1·8	17·2
inclination of axis, with respect to axis of Earth's orbit (deg)	7	<28	3	23	23†	24	3	27	82	29	?
number of moons known	—	0	0	1	—	2	14	10	5	2	1?
atmosphere, chief constituents		none	CO_2	N_2, O_2	none	CO_2	H_2, He	H_2, He	H_2, He, CH_4	H_2, He, CH_4	none?
magnetic field, dipole moment‡ (Earth=1)	3×10^6	$6·6 \times 10^{-4}$	$< 10^{-4}$	1	$< 2 \times 10^{-6}$	3×10^{-4}	$1·9 \times 10^4$	?	?	?	?

* Minus sign denotes rotation is retrograde, i.e. opposite to majority direction.
† That is, orbit is in plane of Earth's equator.
‡ That is, strength of equivalent bar magnet (but some planetary fields are poorly represented by a dipole).

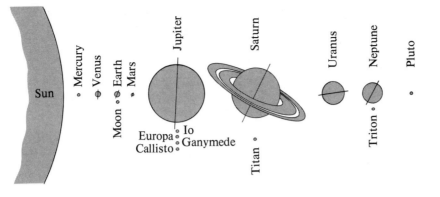

Figure 4.1 Relative sizes of the Sun, planets and moons. Only the seven largest moons are shown, as the others would be mere dots. The axes of rotation, where known, are shown. The separations of the bodies are *not* to scale.

In the same style of thumbnail sketch, the Earth has the highest average density of any planet and is notable for its unusual atmosphere – composed of nitrogen and oxygen – and large amounts of liquid water (see Ch. 10 for how this came about). The absence of all but a few impact craters is attributable to surface geological activity, such as erosion, and, by the same token, the existence of mountains shows it to be currently an internally dynamic planet.

The Moon is large for a satellite, being nearly as large as the largest of Jupiter's moons. Its surface is severely cratered and has light-coloured highlands and dark, fairly flat **maria**.

Mars is smaller and less dense than the Earth. It has a thin atmosphere, mostly of carbon dioxide, which produces thin polar ice-caps plus a trace of water. Its surface is cratered due to impacts, but it also shows many signs of having been an active planet: enormous channels, canyons and volcanoes have been discovered by space probes.

The four inner 'terrestrial' planets and the Moon all have a higher density than the outer planets, and they are thought to be composed chiefly of iron, silicon, oxygen and other relatively heavy elements, with little of the hydrogen and helium that characterise the Sun. They will be discussed in more detail in Section 5.4. Between the terrestrial planets and the outer 'major' ones there is a discontinuity: a great increase in mass and distance from the Sun, and a decrease in density (see Table 4.1).

The asteroids are a large group of minor planets mostly found in the 'gap' between Mars and Jupiter, though some come occasionally within the Earth's orbit. It was once thought that they were the remains of a planet which moved in the 'gap' between Mars and Jupiter, but there is no convincing evidence for this idea. The largest asteroid, Ceres, is only 760 km in diameter and many of the other 2000 known are merely irregular blocks with a total mass far less than any planet. The importance of the asteroids is that they are probably the parent bodies from which meteorites derive; in turn, meteorites play a large part in our understanding of the Earth's composition, as will be discussed in Section 4.5.

Jupiter has a mass greater than that of the other planets added together but, even so, has less than 0·1% of the Sun's mass. It has a low density, which can only be accounted for by a large proportion of hydrogen and helium. Its fast rotation, only 10 h, gives it a pronounced equatorial bulge. Jupiter has 14 known moons with widely different characteristics.

45

Saturn is not unlike Jupiter. Its density is lower – less than that of water – probably because, being smaller, it is less compressed. Its most distinctive feature is its beautiful ring system, consisting of a vast number of separate particles, probably composed of ice and frozen gases, moving in individual orbits without collision.

Uranus and Neptune are about the same size, with densities greater than Saturn's, despite their smaller size, probably due to a lower proportion of hydrogen and helium. Not much is known of them, and it has been discovered only recently that Uranus has rings, though they are far less conspicuous than Saturn's. (Jupiter, too, has very thin rings.)

Pluto is the most distant planet and little is known of it. Only recently a satellite has been discovered, allowing Pluto's mass to be estimated more accurately (Table 4.1). Its orbit is unusual in being inclined at $17°$ to the mean plane of the Solar System, and is so eccentric that sometimes it comes within Neptune's orbit. Since it is also smaller than Neptune's larger satellite Triton, it has been suggested that Pluto was once a moon of Neptune.

Comets are occasionally conspicuous objects with a bright 'head', and a diffuse 'tail' that stretches across half the night sky. However, their heads are only a few kilometres across and are believed to be an agglomeration of solid particles held together by frozen water and gases; they have been called 'dirty snowballs'. The heat of the Sun evaporates the more volatile parts which give rise to the very diffuse tail. It seems they may disintegrate completely, to give rise to meteor showers when the Earth intersects their orbit. At present, comets contribute little to our understanding of the Solar System.

Before we can judge the merits of any theory of formation of the Solar System, we have to decide what features it is essential to explain. That is, it is necessary to distinguish between those features that would be found in all solar systems formed more or less in the same way, and other features that have arisen by chance and are peculiar to our Solar System. This problem would be much easier if our Solar System could be compared with others but, at present, it is impossible to detect an Earth-sized object orbiting even a nearby star, though Jupiter-sized objects have been claimed for some. Thus, it is not known whether solar systems are common, and therefore must be attributed to a fairly normal process, or are rare, so that an unusual combination of events is required to explain the origin of the Solar System. Because most scientists do not like to invoke 'special explanations' without a good cause, the former is often assumed.

The significant features of the Solar System – which have to be accounted for by any formation theory – seem to be as follows.

(a) The planets have a common plane of revolution about the Sun (to within a few degrees) and this is close to the Sun's equatorial plane, i.e. the great bulk of the rotational motion of the system has a common axis (Table 4.1, lines 8 and 9).
(b) The planetary orbits are nearly circular, the most eccentric being those of the smallest planets, Mercury and Pluto (Table 4.1, line 7). (a) and (b) show that the Solar System is a true family with a common origin, and not a random collection of objects attracted by the Sun.
(c) The bulk of the mass is in the Sun, i.e. there is a huge discrepancy of masses, with the Sun 740 times as massive as the planets combined (Table 4.1, line 2).

(d) The motions of the planets about the Sun account for the bulk ($99\frac{1}{2}\%$) of the angular momentum, even though the Sun has most of the mass. This feature of the Solar System has exerted a profound effect upon theories of its formation.

(e) There is a discontinuity in distance, mass and density between the terrestrial planets (Mercury, Venus, Earth and Mars) and the major planets (Jupiter, Saturn, Uranus and Neptune). In the case of density, this must reflect a marked difference in average composition.

4.3   The formation of the Solar System

4.3.1   *Development of theories*

What follows is not a comprehensive review of theories but an attempt to explain how ideas have evolved. For this purpose, it is useful to divide them into three classes, following McCrae (1963):

Class 1. Theories in which the Sun was fully formed before the planets, and the material of the planets came directly from the Sun or another star. The once widely accepted tidal theories belong here.

Class 2. Theories in which the Sun and planets all formed from a rotating cloud, or **nebula**, as a natural result of the evolution of the cloud in response to gravitational and other forces.

Class 3. Theories in which the Sun was fully formed before the planets, as in Class 1, and the planetary material was drawn from interstellar clouds or other source to form a nebula.

Theories of Class 1. The earliest idea of this type was by Buffon, who suggested, in 1776, that a comet in collision with the Sun knocked out material. We know now that comets are far too small to do this, but the idea was revived in 1878, at a time when 'class 2' theories were running into difficulties, but with a star replacing the comet. Then, in 1916, Jeans showed by calculation that a near approach would be sufficient, huge tides being raised on both the approaching star and the Sun, increasing in amplitude until material is pulled off to form a cigar-shaped filament between the two bodies (Fig. 4.2). The star then returns to the depths of space, leaving the hot filament to condense into planets, with the most massive near the central, fatter part of the cigar.

This theory was widely accepted, but it has several drawbacks. Although the theory seemed able to get material into orbit at a considerable distance from the Sun, and so account for the large angular momentum of the planets about the Sun, it did not account for the high speed of rotation of the planets about their own axes. Several variants of the theory were proposed to overcome this difficulty, but a more general drawback was pointed out in 1939 by Spitzer. This was that, to provide enough material, the filament would have to come from the interior of the Sun or other star where the temperature is millions of degrees, and at such a temperature the filament would disperse before it could cool.

Only one theory of this class is still viable, that by Woolfson (1978a, b). Broadly, he follows Jeans' tidal theory, but the star that approached the Sun was one of many **47** in a newly formed star cluster and was still large and cool, not having contracted to the nuclear fusion stage (Section 4.6.2). By having a cool star, he obviates the difficulties of hot material, and, by making the encounter take place in a cluster, he makes a near approach far more likely than in Jeans' theory. Woolfson has worked out his theory in considerable detail using computer simulation, to account for many features of the Solar System.

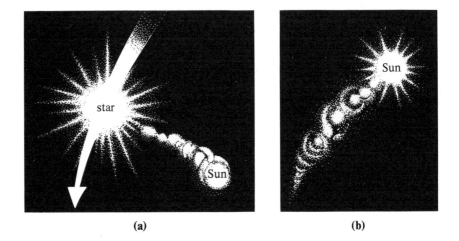

Figure 4.2 Tidal theory of planetary formation. (a) Passage of a star near the Sun raises huge tides and finally draws out a filament of incandescent gaseous material. (b) After the star departs, the filament condenses into the planets, the largest in the middle of the filament where it is thickest.

(a)   **(b)**

Theories of Class 2. The theories of Class 2 are the nebula theories and can be traced back to Kant who proposed, in 1775, that the matter of the Solar System originally was dispersed and came together by gravitational attraction. But many of his details were implausible and it was 1830 before the theory was put on a more scientific foundation. The Marquis de Laplace pointed out that a rotating cloud, contracting under its own gravity, would have to spin faster and faster to conserve angular momentum. Eventually, the centrifugal force at its perimeter would exceed the pull of gravity towards the centre, so that a ring of material would be thrown off. Laplace supposed that the ring would condense into a planet, and that the remaining cloud would continue to contract and repeat the process at intervals (Fig. 4.3).

Figure 4.3 Laplace's nebula theory. A disc of gas spins faster as it contracts, until the centrifugal force exceeds the gravitational attraction at its edge and a ring is thrown off. This is repeated at intervals. The rings form into planets and the central mass becomes the Sun.

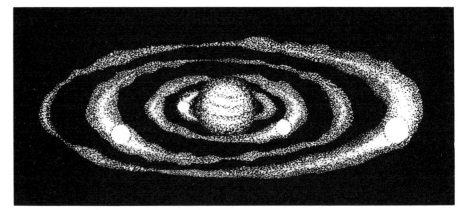

Not until near the end of the 19th century was it realised that, according to this model, the Sun should be rotating near to its centrifugal limit (with a period of a few hours) instead of very much slower (it takes 27 days for one rotation). Thus, it ought to have most of the angular momentum of the Solar System. In 1918, Jeffreys showed that such a gas cloud could not form the planets directly by condensation, because of the way angular momentum and mass are separated in the Solar System. Another unexplained point was why the nebula should throw off material in rings, rather than continuously to form a disc. Because of these difficulties, attention turned to the tidal theories of class 1.

Nebula theories easily account for the planets all rotating in the equatorial plane of the Sun, with the planets and the Sun rotating in the same sense. But, the major problems are (i) to account for the slow rotation of the Sun and (ii) to discover plausible mechanisms whereby the material thrown off would collect into planets. A number of possible solutions have been discovered for each of these difficulties, and some will be presented in Section 4.3.2.

Theories of Class 3. These theories, in many ways, are hybrids of the two previous classes and try to avoid their difficulties, usually by postulating suitable conditions rather than deriving them. The Sun is taken to be already in existence before it acquires a nebula by passage through a cloud of gas and dust, or in an unspecified way.

In the mid-1940s, Von Weizsäcker assumed that the Sun had a gaseous envelope possessing angular momentum. This would flatten into a disc under gravity, but would be unstable due to its relatively high gas pressure. It follows from Kepler's laws that the further a particle is from the Sun, the longer it takes to complete an orbit (see Eq. 3.5), whereas friction in the gas would tend to make all the particles orbit in the same time. Von Weizsäcker concluded that the inner parts of the disc would slow and fall into the Sun, slowing its rotation, while the outer parts would break up into a regular system of vortices which would condense to form the planets.

However, the regular pattern of vortices is very unlikely, and later theories on these lines replaced it by random turbulence, which prevents the disc becoming unstable as above. Ter Haar (1948, 1950) supposed, therefore, that the material would gather itself together to form planets by some process of **accretion**, i.e. the 'sticking together' of solid particles. He explained the difference of density between the terrestrial and the major planets as due to increasing temperature in the nebula towards the Sun. As a result, only the least volatile elements would be able to form solid particles near the Sun, and, as such elements are also dense, the planets formed would be dense too. The importance of a temperature gradient in the Solar Nebula is widely accepted today (Sections 5.2 and 5.5).

A totally different type of theory depends upon forces upon ions – atoms that have lost or gained one or more electrons and so become electrically charged – in a magnetic field. These ideas were especially developed by Alfvén who proposed, in 1954, that interstellar matter fell into the Sun and that the resulting high velocities sometimes produced ions when the constituent particles collided. Once ionised, the motions of the particles would be influenced by magnetic fields which are commonly present. Because the various gases and solid particles differ in their ease of ionisation, Alfvén used this mechanism to explain the difference in composition between the terrestrial and major planets.

None of these theories, when considered quantitatively, is able to give a plausible account of the features of the Solar System without relying on arbitrary postulates, and no theories of this class are proposed today. However, some of their ideas have been incorporated into current theories.

4.3.2 *Modern nebula theories*

At present, there is no theory that can satisfactorily account for *all* the features of the Solar System, starting from a plausible initial state. On the other hand, any given feature can be explained in at least two ways, according to different theories.

Of the three classes of theory discussed in the previous subsection, theories of class 3 have no current versions, while class 1 is represented only by Woolfson's tidal theory. The great bulk of effort is going into nebula theories, which exist in a number of versions, some of which will be summarised below. All these theories, and Woolfson's too, do not consider the evolution of the Solar System in isolation but as part of the evolution of a cluster of stars which forms from an interstellar cloud. Evolution has to go through a number of stages which will be considered in turn.

Formation of protostars from an interstellar cloud. The Sun is only one of about 10^{11} stars which belong to the Galaxy (Figs 4.4 and 4.5), most of which are concentrated into the galactic centre and the spiral arms. It is believed that the spiral arms are not fixed features but are 'waves' of higher density that sweep around the centre. Stars travel more slowly when they reach the arms and this causes the increase of density, just as vehicles close up when a line of traffic is forced to slow down.

Figure 4.4 The Galaxy in (a) elevation and (b) plan. Most of the matter is in the centre and in the spiral arms but, in addition, there are globular star clusters containing many stars, forming a spherical 'halo'. The approximate position of the Sun is shown by a cross.

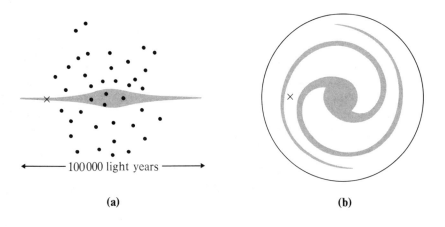

100 000 light years

(a)

(b)

(a)

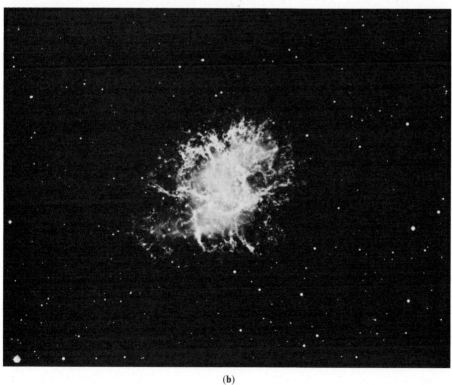

(b)

Figure 4.5 (a) Spiral galaxy in Ursa Major: this is similar to our own Galaxy, a few of whose stars are dotted over the picture. (b) The Crab Nebula, the expanding remains of a supernova that exploded in AD 1054 (see Section 4.6). (Photographs from the Hale Observatories, reproduced by permission of the Royal Astronomical Society, London.)

51

Between the stars is the interstellar medium, of gas, plus a little dust which consists of solid grains a few microns (μm) across. Within the medium are interstellar clouds which are both cooler and denser: a typical cloud has a density of 10 atoms per cubic centimetre, a temperature of 100 K ($-170°$C) and a total mass of 10^4 suns. They do not have sufficient density to contract under their own gravity, and are prevented from expanding by the surrounding interstellar medium. However, when one enters a spiral arm, the resulting compression may be sufficient to initiate self-contraction. This idea is supported because large bright stars and pre-supernova stars, which are young stars (see Section 4.6.2), are found mostly along the leading edges of the spiral arms.

A problem in forming stars from such a cloud is that if each star had a proportionate share of the angular momentum of the cloud it would rotate far too fast for stability. That is, its spin period would have to be reduced 10 000 times before material would cease to be flung off at the equator. And the Sun rotates a hundred times more slowly than this!

One solution proposed for this problem relies upon the cloud being highly turbulent. Eddies permit temporary cloudlets or 'floccules' to form (McCrea 1978). If two of these, with opposite rotations, collide they may make a more stable aggregation which can attract further floccules, the whole having proportionately little angular momentum. As a result, the angular momentum of the original cloud is concentrated into the relative motions of the stars formed from it, rather than into their rotations.

Another solution (Reeves 1978) relies upon clouds being partly ionised, due to starlight and cosmic rays. Because collisions in the extremely thin gas are few, the ions will behave as a near-perfect electrical conductor. Weak magnetic fields are known to be present in interstellar clouds – typically 3×10^{-10} T (tesla), or 10^{-5} of the Earth's field – and extend into the surrounding interstellar medium. These magnetic fields interact with the ionised cloud and effectively stiffen it (see Note 4). As a result, when the cloud tries to rotate faster as it contracts, the magnetic interaction acts as a brake, transferring angular momentum from the cloud into the surrounding medium.

Regardless of the detailed mechanism, it is evident that, when a cloud forms into stars, its angular momentum is not transferred to them for, if it were, they would rotate far too quickly to be stable. Moreover, many of the stars are double, or binaries, and the axes of rotation of the pairs do not exhibit a preferred direction, as would be expected if their rotational motion derived from the rotation of the cloud by simple sharing.

Evolution of a protostar into protoplanets around a central mass. So far, the cloud has divided into many thousand separate smaller clouds, or protostars. These will be much denser than the parent cloud and so can hold themselves together by their own gravity. Because the protostar has contracted so much, it will be rotating much faster than the parent cloud, even though it does not have a proportionate amount of its angular momentum. The further collapse of the protostar is opposed by gas pressure, centrifugal force and, according to some, its magnetic field. Thus, the protostar will be a rough replica of the parent interstellar cloud, but with different values of density, size and other parameters.

As a result of the faster rotation, the protostar will have flattened into a disc, or solar nebula. This can be understood with the aid of Figure 4.6. Gravitational attraction is everywhere towards the centre of a mass, but centrifugal force is directed perpendicularly to the axis of rotation. The two forces are not antiparallel (except in the median plane), and so the combined force will move the gas and dust of the protostar nearer to the median plane as the protostar contracts and rotates more quickly. The next problem is how the solar nebula can throw off some material that will form the planets, though the bulk of it contracts to become a sun. According to Prentice (1978), the nebula has a diameter many times that of the Solar System and a density of about 10^{-16} kg m^{-3} (about 10^{10} atoms per cubic centimetre). It is still cool, about 100 K, because although heat has been released by the contraction, it has been lost by radiation. However, at this point in its evolution, the nebula has just become dense enough to prevent radiation escaping directly and, therefore, any further contraction causes the nebula to heat up. This heat tends to raise the temperature and hence the gas pressure, and so helps to stabilise the nebula against further contraction. The nebula enters a phase of quasi-equilibrium, in which heat is lost only slowly by leakage from the interior to the surface, whence it is radiated into space. This causes the nebula to contract slowly and *to heat up*.

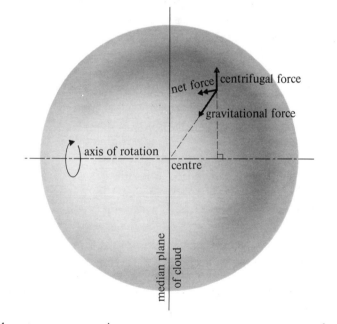

Figure 4.6 Contraction of a spherical cloud into a disc. Because the gravitational and centrifugal forces are not antiparallel, there is a net force towards the median plane, which ensures that the mass flattens towards that plane as it contracts.

In the laboratory, compressing a gas at constant temperature not only causes it to contract but also raises its pressure, so that there comes a point when the internal gas pressure balances the applied pressure. But in a large self-contracting mass of gas this simple mechanism for balance does not operate. This is because the contraction, by bringing all the matter closer together, increases the gravitational force, and this increases faster than the increase in pressure due to the contraction. The only way balance can be achieved is for some of the heat released by the contraction to remain within the cloud, raising its temperature and hence the gas pressure. Hence the seeming paradox that heat loss causes the cloud to *heat up*, as well as contract, with part of the energy released going to make good the external heat loss and part to

53

increase the internal temperature. This behaviour plays a central role in stellar evolution (Section 4.6.2).

This phase of slow evolution ends when contraction has raised the temperature high enough to evaporate the grains of solid gases (CO_2, NH_3, etc.), for this absorbs heat and so prevents the compensatory rise of temperature described above. The material of the nebula begins to fall towards the centre and stability will not be regained until all demands of latent heat – not only vaporisation, but also dissociation of gas molecules such as H_2 into atoms, and their ionisation – have been satisfied. When this happens, its diameter is only a few times that of the present Sun, and it has a temperature of thousands of degrees.

This collapse takes place only in the innermost few per cent of the nebula, because this is the part that first reaches the temperature of vaporisation. It collapses and regains stability as a central 'core', before the outer part has contracted far. The outer part continues to contract, but reaches stability due to a rise in temperature produced by both its contraction and by heat from the hot central 'core'. Prentice believes that the nebula will have a diameter of about that of Neptune's orbit, and be highly turbulent at this stage.

The turbulence is important because, by exchanging material between inner and outer parts of the nebula, it prevents the inner parts rotating faster than the outer parts, as otherwise would be the case in a contracting nebula. This transfers angular momentum from inside the nebula to a non-turbulent annulus on the edge of the nebula. Finally, the annulus is left behind as a ring as the nebula continues to contract. This process is repeated at intervals, producing a series of rings of roughly the same mass but progressively closer together as the central core is approached. Thus, according to this theory, Laplace was correct when he assumed that rings would be produced. The formation of rings corresponding to the existing planets would take only about 0·2 Ma, though the final contraction of the nebula onto the core, to form a sun, would take several million years.

Alfvén (1978) believes that magnetic fields have a vital role to play in the development of the nebula, once it is partly ionised. Electromagnetic forces (Note 4) transfer angular momentum outwards, so permitting a contraction of the bulk of the material and also produce a partial separation of the elements.

Other theories also rely upon magnetic or turbulent viscous forces to permit the nebula to collapse, but differ in detail and will not be considered further.

Formation of the planets. Evolution of the nebula has concentrated most of its mass into a small hot central volume that is not yet quite a sun, but with some material left behind at planetary distances, in a disc or rings. How did this material collect together to form planets? There are various ideas, but all agree it involved a number of stages.

In a cloud, every element whose melting point is less than the temperature of the cloud will exist as a vapour or gas. However, those whose melting points are above the cloud's temperature will exist only partly as solid grains. This is because all materials have some vapour pressure, though it will be very small if the temperature is well below its melting point. In an interstellar cloud the pressure is so low that a significant fraction of even iron and the silicates exists as a vapour (at these low pressures there is no liquid phase).

When a solar nebula forms from part of the interstellar cloud, the increase in pressure causes some of the vapour to condense into grains. At the same time, the gravitational attraction of the matter in the nebula attracts the grains towards the median plane because, though centrifugal force prevents them moving towards the axis of the nebula, it cannot prevent movement perpendicular to the plane of the nebula (see Fig. 4.6). At first, the grains fall slowly because of gas resistance, but they grow by condensation of vapour and so fall progressively faster. If the vapour is not exhausted, they will reach the median plane in 10 years, with a diameter of 3 cm (Goldreich & Ward 1973).

If the total mass in the unit area (i.e. the surface density) of the dust disc is great enough locally, chance concentrations of matter in the disc will contract under their own gravity. As a result, the dust disc concentrates into local aggregations of material, called **planetisimals**, perhaps 100 m across. In turn, the distribution of these planetisimals is also unstable and they aggregate into second-generation planetisimals about 5 km across, within a few thousand years (Goldreich & Ward 1973). Thereafter, growth is probably by collisions as well as by gravity, with planets forming in about a tenth of a million years. Other theories depend upon grains sticking together to initiate the accretion process. Kerridge (1978) lists several mechanisms, which include impact welding and electrostatic attraction. Magnetic attraction between iron grains has also been suggested. Once a planetisimal has grown large enough, gravitational attraction becomes important for adding material.

It is agreed that sticking mechanisms would result in accretion times of tens of millions of years, much longer than by the gravitational instability mechanism described above. They would, therefore, be unimportant unless there were turbulence in the accreting disc, which would disrupt the growth of planetisimals.

In other theories, nebula material is gathered into discrete clouds, or protoplanets, by gravitational forces, and only then do some of the above accretion mechanisms cause these protoplanets to form the planets we know.

Fractionation of the elements, and the role of temperature. During the process of producing a sun and planets from part of an interstellar cloud, there are many opportunities for some degree of segregation, or **fractionation**, of the elmenets. One of the chief causes is the difference of the grains, which are composed of the less volatile and usually heavier elements, from the gases. Fractionation could occur by simple settling of the grains through the gas, as was seen in the formation of a dust disc; because grains are far less likely than gases to acquire an electric charge and so be affected by electromagnetic forces; or because of the ability of grains to stick together by the various mechanisms described. Silicate grains are also likely to separate from metal grains, because of their different densities, and experiments have shown that metal grains are much more likely than silicate grains to adhere after a collision. Also, if magnetic attraction between grains is important then, clearly, this favours accretion of iron and nickel.

Temperature has hardly been mentioned, so far, but it plays a critical role in most theories of planetary accretion and composition. What elements or compounds form grains will depend, amongst other factors, on temperature, for all materials can be vaporised given a sufficiently high temperature. It is also agreed that the

55

nebula will heat up due to its contraction, with the highest temperature near the centre. However, theories differ as to whether temperature exerts its greatest effects during the early stages of evolution of the nebula, when it is heating up, or later, as the protoplanets begin to cool and grains reform by condensation.

Prentice (1978), in his theory, suggests that, though rings are formed within Mercury's orbit, the temperature there is too high for any grains to exist and so accretion cannot begin. At Mercury's orbit, the temperature is about 1000°C and only the most refractory materials – some metals and silicates – can condense, accounting for its high density. This is still broadly true for all the terrestrial planets, and Jupiter is the innermost planet that can accrete the more volatile and lighter elements, such as compounds of H, C, N and O (see also Section 5.2).

Some of these fractionation mechanisms are likely to produce differences between the bulk compositions of the planets, particularly between the terrestrial and the major planets, whilst others could produce layering within a planet, by accreting some elements before others.

What has been learned from this enquiry into the formation of the Solar System? Though there is still much uncertainty, some broad conclusions can be drawn. The Solar System probably derived ultimately from an interstellar cloud, along with many other stars. The composition closest to that of the original Solar Nebula is to be found in the Sun, because it formed without most of the complicated processes of accretion and fractionation that the planets experienced (many of which were influenced by the evolving Sun). The planetary masses and compositions can be derived from roughly equal masses of solar composition, by progressive fractionation. Jupiter has lost relatively little hydrogen and helium, Saturn rather more, while Uranus and Neptune have lost most of these elements and are made up chiefly of compounds of carbon, nitrogen and oxygen, with a little hydrogen. The terrestrial planets are composed predominantly of heavier elements. As yet, a more detailed knowledge of compositional differences cannot be deduced from theories of formation of the Solar System, and a different approach will be used in Chapter 5.

4.4 Solar abundances of the elements

Solar abundances are determined from spectral absorption lines produced when light from the visible surface of the Sun passes through the thin, cooler layers above. The wavelengths absorbed are those which would be emitted if a vapour of the element were heated, or otherwise excited. An example is the yellow light of sodium streetlights. The absorption lines are seen as dark lines in the spectrum when sunlight is split into its component wavelengths by a spectrometer.

The strength of an absorption line depends not only upon the amount of the element present but also upon the temperature, the pressure and the efficiency of a particular element in absorbing light ('oscillator strength'). This last factor is difficult to determine by calculation or experiment, and so there are uncertainties of up to a factor of 10 or more for the abundances of some elements.

Since the Sun derives its energy from nuclear reactions which convert one element into another, there is the question whether the abundances measured have been affected by these processes since the Sun formed. However, from theories of stellar evolution (Section 4.6.2), it is believed that these processes are confined to the deep interior, with the exception of those affecting lithium, so that the outer layers retain the original composition.

A knowledge of the solar abundances provides the starting composition of the Solar Nebula, and this can be compared with the abundances found elsewhere in the Solar System, in particular in the meteorites and in the Earth's crust. This is done in Figures 4.7 and 5.4, where the abundances have all been compared to that of silicon.

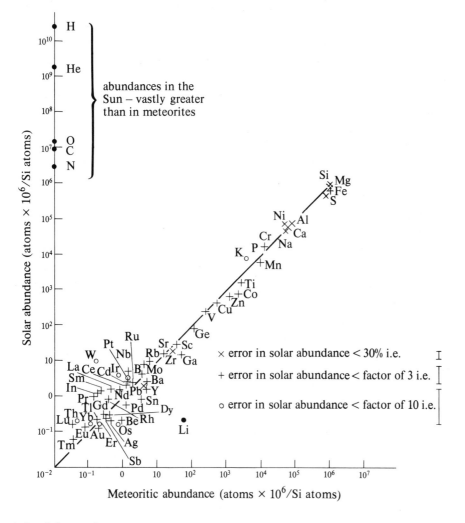

Figure 4.7 Comparison of solar and chondritic meteorite element abundances (silicon is set at 10^6). H, He, O, C and N are not compared because they have largely been lost from meteorites, owing to their great volatility, but are shown on the solar abundance axis to demonstrate their preponderance in the Sun. The anomalous position of lithium (Li) is discussed in Section 4.6. (Data from Trimble 1975.)

4.5   Meteorites

4.5.1   *Introduction*

Meteorites play a vital role because it is believed that they represent various stages in the planetary accretion process – fossils of the evolving early Solar System, as it were – and so provide evidence of accretion processes that, in the planets, have been obliterated by subsequent processes. In particular, it is thought that some of them may be equivalent to the bulk composition of the Earth – subject to modification by some loss of elements – and the idea has led to the Chondritic Earth Model which will be developed in Section 5.4.

By definition, a meteorite is a piece of extra-terrestrial matter that survives passage through the Earth's atmosphere to land on its surface. Meteors, on the other hand, vaporise completely, giving rise to 'shooting stars'. As well as being much smaller than meteorites, they are believed by most scientists to be quite separate, and they will not be considered further.

The sudden arrival of a meteorite, often accompanied by an impressive fireball, sometimes led to them being regarded with reverence; for instance, the sacred stone of the Islamic world, in the Kaaba, Mecca, is reputed to be one. Elsewhere, tribes have put metal meteorites to more practical use, though beating them into swords rather than ploughshares. The fanciful stories surrounding meteorites perhaps provoked scientists to undue scepticism about their origin, so that Chladni was reluctant, in 1794, to publish his reasons for believing them to be extra-terrestrial matter. However, his theory soon was widely accepted. (A notable disbeliever was Thomas Jefferson, who helped draft the American constitution and later became President. He remarked of a fall reported by two professors of Yale: 'It is easier to believe that Yankee professors would lie, than that stones would fall from heaven.' But, as Jefferson lived in Virginia, this may tell us more about his views on Yankees than on meteorites.)

Meteorites are often divided into four major groups (Table 4.2). **Irons** (Fig. 4.8a) contain over 90% of nickel and iron as a metallic alloy; **stones** are predominantly of silicate minerals, mostly of types common on Earth; while **stony-irons** are usually about half metal and half silicate (Fig. 4.8b). The stones are then divided into **chondrites**, which contain **chondrules** (Fig. 4.8c), and **achondrites**, in which chondrules are absent or very rare, and which, unlike the other three types, contain little free metal ($< 1\%$). Irons are the most common in museum collections, but this is because they are more likely to be recognised, being more obviously different from terrestrial rocks. Table 4.2 shows that chondrites are really far more common. However, we don't know how well this table reflects the abundances at source, if only because most known meteorites have only landed in the past few hundred years.

Table 4.2 Classification of meteorites. (Based on data from Wasson 1974.)

				Observed falls	
				(%)	(No.)
differentiated meteorites	{ irons			1.1	8
	{ stony-irons			3.2	22
	{ achondrites }		stones {	8.3	57
undifferentiated meteorites	chondrites }		{	87.4	602

Meteorites may be further divided into many classes, using various criteria, but usually with emphasis upon the predominant minerals. For our purposes, it is more useful to divide meteorites first into two groups: **differentiated meteorites** (Table 4.2), which contain evidence of post-accretional melting (and other processes) that must have led to a marked segregation of the elements, and **undifferentiated meteorites**, which have not been melted and so are much closer to the original composition from which their parent bodies and the whole Solar System formed.

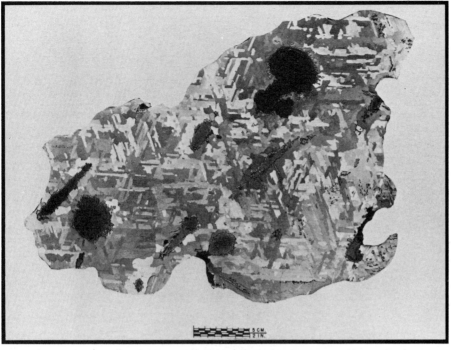

(a)

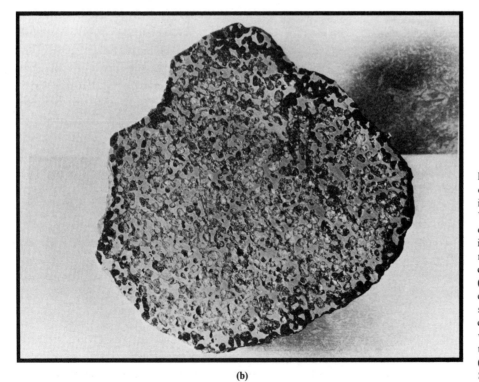

(b)

Figure 4.8 Polished sections of meteorites. (a) Iron: the intersecting lines are Widmanstätten patterns; the dark areas are silicate inclusions. (b) Stony-iron: the metal matrix encloses olivine crystals. (c) Chondrite (Allende carbonaceous chondrite): the many small spheres of different sizes are chondrules; the two prominent white masses are 'high-temperature inclusions'. (Photographs from the Smithsonian Institution.)

59

(c)

4.5.2 *Undifferentiated meteorites: chondrites*

Chondrites are characterised by the presence of chondrules (Fig. 4.8c) which have roughly the same composition as the matrix in which they are embedded. Apparently, chondrules acquired their near-spherical shape by being melted prior to being incorporated in the matrix, though it is not known exactly how or when they formed. One suggestion is direct condensation from a vapour, but this would require improbably high pressures. Another is impact fusion of grains or small planetisimals, for which support comes from at least one known example of partial impact melting. This would have happened at an early stage of accretion of the material of the Solar Nebula into the meteorite parent bodies.

Chondrites may be classified in two ways. One is by composition, and Figure 4.9 illustrates how they may be divided into discontinuous classes, in this case using ratios between iron, oxygen and silicon. These classes are thought to reflect different parent bodies in different states of oxidation, and with different Fe/Si ratios (discussed in detail in Section 5.4). The second classification is a continuous one and is based upon the degree of recrystallisation or 'metamorphism' that a meteorite has experienced. For instance, in some meteorites the chondrules are very distinct, in others less so, whilst in some they are discernible only as 'ghosts', with crystals continuous across their boundaries. Expressed more generally, the meteorites in the lowest grades of metamorphism are furthest from chemical equilibrium; that is, if heated sufficiently some of the minerals could not coexist together and their elements would rearrange themselves, e.g.

$$Fe + 4MgSiO_3 + Fe_3O_4 \rightarrow 4MgFeSiO_4$$

iron   pyroxene   magnetite   olivine

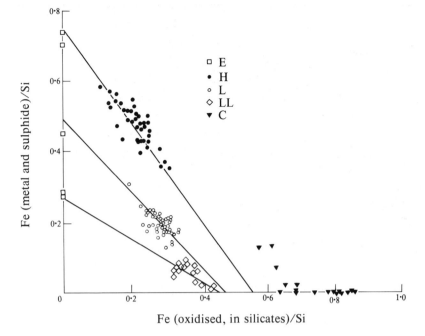

Fe (oxidised, in silicates)/Si

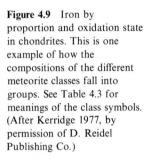

Figure 4.9 Iron by proportion and oxidation state in chondrites. This is one example of how the compositions of the different meteorite classes fall into groups. See Table 4.3 for meanings of the class symbols. (After Kerridge 1977, by permission of D. Reidel Publishing Co.)

This has happened in chondrites of higher metamorphic grades.

Van Schmus and Wood in 1967 devised a classification of chondrites based upon both composition and degree of metamorphism (Table 4.3). Meteorites along a single row represent a single chemically related series, with increasing metamorphism to the right. Consequently, those to the left, being least modified, are regarded as the most 'primitive' material. The carbonaceous chondrites differ from other types in having significant amounts of water (combined as water of crystallisation), carbon compounds and sulphur, but little uncombined metal. Figure 4.9 shows that they are the most oxidised group of meteorites and have a high Fe/Si ratio. The car-

Table 4.3   Classification of chondrites.

Chondrite type	Degree of metamorphism					
	increasing degree of recrystallisation → increasing degree of chemical equilibrium → decreasing amount of volatiles, etc. →					
	1	2	3	4	5	6
enstatite, E*			E3 (0)	E4 (3)	E5 (2)	E6 (6)
high iron, H			H3 (6)	H4 (23)	H5 (53)	H6 (32)
low iron, L			L1 (9)	L2 (11)	L3 (28)	L4 (117)
very low iron, LL			LL1 (6)	LL2 (1)	LL3 (7)	LL4 (20)
carbonaceous, C	C1 (5)	C2 (18)	C3 (9)	C4 (1)		

Numbers in parentheses show number of falls in the category. Those in the rectangular box form the *ordinary chondrites* which are the most abundant types. (Modified from Van Schmus & Wood 1967.)
*Contains the Mg-rich orthopyroxene 'enstatite'.

61

bonaceous chondrites have experienced the least metamorphism of any meteorites, for they would have lost their more volatile components if heated to only 180°C, whereas meteorites in column 6 have experienced a temperature of 700–800°C.

The significance of carbonaceous chondrites is that, being the most 'primitive' of meteorites, they are believed to be closest in composition to the original Solar Nebula. A comparison of chondritic and solar abundances (Fig. 4.7) shows marked similarities in the relative abundances of most elements, except the most volatile. An exception is Li, which is probably destroyed within the Sun by nuclear reactions (Section 4.6.2). This agreement confirms the relevance of meteorites to an understanding of the Solar System. Meteorites also provide information on isotopic ratios and on physical and chemical conditions in the evolving Solar Nebula, information not available from studying the Sun.

4.5.3 *Differentiated meteorites: irons, stony-irons and achondrites*

Irons. Most irons, when sectioned, polished and etched with acid, reveal a trellis-like figuring called Widmanstätten patterns (Fig. 4.8a). Such patterns arise when two minerals crystallising cease to be fully miscible in solid form as the temperature is lowered. Suppose the atoms of two elements are similar but not identical – such as iron and nickel – so that each separately would form crystal lattices that differ slightly. The two types of atom are freely interchangeable in a crystal at high temperature, because of the looseness of the expanded lattice. But, as the temperature is lowered, the differences between the two types of atom become noticeable. There comes a point when the energy of the whole system can be lowered by segregating the atoms into two separate lattices, one rich in and the other poor in a given element, even though there is some mismatch where the lattices abut. To minimise this mismatch, the newly formed lattices grow along preferred directions in the host lattice, as **exsolution lamellae**. An example familiar to petrologists is the perthite structure in alkali feldspars.

Figure 4.10 shows the phase diagram for solid nickel–iron mixtures. Consider a mixture of, say, 10% nickel in iron, initially at 1000°C (point A in the figure). At this temperature, the two elements are fully miscible in solid solution, but when the temperature has fallen to B this ceases to be true. Below B, kamacite (γ nickel–iron) of composition B_1 forms within the host lattice of taenite (α nickel–iron). Further cooling to C increases the disparity of the two lattice compositions, though the proportions of C_1 and C_2 must be such that the bulk composition is 10% Ni, 90% Fe. The kamacite forms along certain planes in the taenite which correspond to the surfaces of an octahedron; hence the name 'octahedrite' is often used for these meteorites. The surfaces of an octahedron (two pyramids base to base) belong to only four planes, since opposite faces are parallel, and random sections through the crystal will give rise to different Widmanstätten patterns, though generally like those of Figure 4.8a and the inset to Figure 4.10.

Full development of exsolution lamellae requires that the atoms are given sufficient time to rearrange themselves by diffusion through the solid, and, since diffusion becomes slower as the temperature decreases, eventually the composition of the lattices becomes 'frozen'. The faster the cooling rate the higher the temperature at which this happens. Detailed examination of the composition of the exsolution lamellae give values of 1 to 10°C in a million years for the cooling rates of a range of iron meteorites. These cooling rates are most plausibly explained if the meteorite had been part of a hot body that cooled slowly because of its size, plus the insulating effect of a 'mantle' of silicates. Calculations show that the required diameter would be·a few hundred kilometres, which is comparable with the size of the larger asteroids.

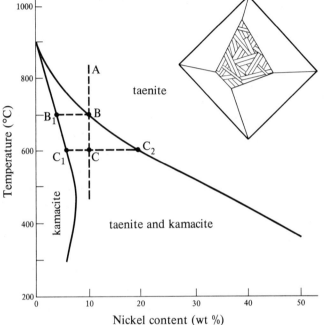

Figure 4.10 Phase diagram of solid Fe–Ni, at atmospheric pressure. Kamacite forms along octahedral planes of the original taenite, and the inset shows the Widmanstätten patterns resulting from an arbitrary slice through the lattice. (After Goldstein and Short 1967.)

Achondrites. Of the four groups of meteorites, only the achondrites are at all similar to known terrestrial rocks. Some are close to olivine-rich basalts and fine-grained peridotites and, therefore, could only have been formed by chemical-refinement processes involving melting within a body, as described in Section 5.2.2.

Stony-irons. The most abundant meteorites of this group have a nickel–iron matrix with inclusions of olivine and minor amounts of other silicate minerals (Fig. 4.8b). The metal phase is similar to that in the octahedrites and the two meteorite types are probably related.

   The minerals of all three types of differentiated meteorites crystallised at high temperatures and, in the case of the irons, there is evidence of slow cooling. Probably they are fragments produced by collisions between asteroids. Support for this comes from the highly irregular shapes of the smaller asteroids, and from exposure ages which are deduced from the amount of alteration produced by cosmic-ray bombardment only on exposed surfaces. The exposure ages are much smaller than the formation ages (see Section 4.5.4), and some are only a few million years. Often, the exposure ages of a given class of meteorite are the same, suggesting they all come from a single collision.

It is thought that the differentiated meteorites probably derive from once-molten asteroids, in which the metal settled to form a core, surrounded by a silicate mantle and with a transition layer between. It cooled slowly and later was broken apart by a collision, yielding irons, stony irons and achondrites from the three layers. The relevance to the Earth's layers should be obvious.

4.5.4 *Meteorite ages and formation intervals*

The formation ages of meteorites – the time when they acquired their present mineralogical composition – have been determined by the rubidium–strontium isochron method (Note 5) and in other ways. Most of their ages are in the range 4300 to 4700 Ma old. This is close to the age of the Solar System, which is determined from lead isotopic evidence (Note 6).

Additional evidence that meteorites attained their present state early in the evolution of the Solar System comes from estimates of **formation intervals**; that is, the time between the production of certain radioactive isotopes and their incorporation in the meteorites. It is believed that these isotopes (see Note 7 for an explanation of this and related terms) were produced in supernovae shortly before the formation of the Solar Nebula (Section 4.6.2). The ones of chief interest to us are those with short half-lives:

^{244}Pu → fission fragments, including some xenon isotopes   half-life 82 Ma
^{129}I → ^{129}Xe half-life 16·4 Ma
^{26}Al → ^{26}Mg half-life 0·72 Ma

If any of the parent isotopes were incorporated in a mineral before they had decayed to negligible amounts, their subsequent decay would produce a quantity of daughter isotope that may be detectable as an isotope anomaly. The method will be illustrated by considering ^{26}Al.

Aluminium has only one stable isotope, ^{27}Al, which would be synthesised at the same time as ^{26}Al (Section 4.6), so if any aluminium were incorporated into a mineral it would contain both ^{27}Al and ^{26}Al in the ratio of that time. Consider several different minerals that formed at the same time, but which incorporated Al and Mg in different proportions. Suppose one mineral contained no Al: it would have $^{26}Mg/^{24}Mg$ in the ratio of the time, and this ratio would not change subsequently, because it had no radioactive ^{26}Al. On a diagram of $^{26}Mg/^{24}Mg$ versus $^{26}Al/^{24}Mg$, it would be plotted at A on Figure 4.11a. Suppose a second mineral contained both Al and Mg: at the time of its formation it would plot at B_1, level with A. But because it contained ^{26}Al (not shown) as well as ^{27}Al, its $^{26}Mg/^{24}Mg$ ratio would increase with time, reaching a final value, B_2, when effectively all the ^{26}Al had decayed away. Other minerals, with different Al/Mg ratios plot elsewhere on the same straight line. If, instead, all the minerals had formed after all the ^{26}Al had decayed away, they would retain their initial $^{26}Mg/^{24}Mg$ ratio and plot as a horizontal straight line. Thus, the steeper the line the sooner after its formation the aluminium must have been incorporated into the mineral. The ratio in which ^{27}Al and ^{26}Al are synthesised can be estimated from

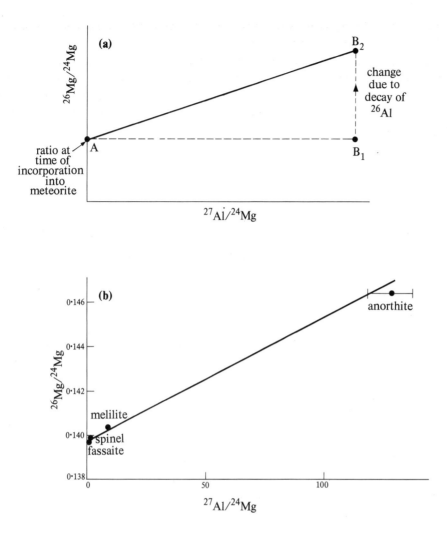

Figure 4.11 Evidence for extinct ^{26}Al. (a) This shows schematically how the compositions of minerals with different Al/Mg ratios evolve by decay of ^{26}Al. (b) This shows results for a single inclusion from the Allende carbonaceous chondrite; for further details see text. (Data from Lee *et al.* 1978, by permission of Pergamon Press.)

nucleosynthesis theory (Section 4.6.1), and then the slope of the line can be used to calculate a formation interval: a steep line indicates a short formation interval.

Figure 4.11b shows results for different minerals from a single inclusion in the Allende meteorite, a C3 carbonaceous chondrite, that fell in Mexico in 1969 (Fig. 4.8c). Since the half-life of ^{26}Al is only 0·7 Ma, it is obvious that the formation interval must have been short, and calculation gives a maximum period of 3 Ma. These inclusions are composed of the most refractory minerals indicating that they are the first minerals to condense from a cooling vapour, and the probable interpretation is that they are very early condensates of the Solar Nebula (Black 1978).

This result raises a number of points. First, such a short interval indicates that the formation of the Solar System is related to the synthesis of the elements, and it has been suggested that a supernova both synthesised some of the elements and triggered the formation of the Solar System (see Sections 4.3.2 and 4.6.1). Secondly, the incorporation of even small amounts of short-lived radioactive isotopes, particularly ^{26}Al, into the newly formed planets and asteroids would have provided a source of heat that might have helped to melt them (see Section 5.2.2).

The short formation interval based upon ^{26}Al contradicts estimates of about 100 Ma based upon ^{244}Pu and ^{129}I. This embarrassment is removed if these two other radioactive isotopes were formed by an earlier supernova when the interstellar cloud, from which the Sun subsequently formed, made an earlier passage through a spiral arm, which is likely to have occurred about 100 Ma before (Section 4.3.2).

Though much has yet to be understood about the meteorites, the following conclusions seem to be justified.

(a) The fabric of chondrites is quite unlike that of any terrestrial rock and required very different conditions in which to form. These are identified with early stages in the development of the Solar Nebula.

(b) The carbonaceous chondrites are a close approximation to the material of the Solar Nebula, having lost only the most volatile elements. It is, therefore, plausible to regard them as a starting point from which the composition of the Earth has evolved. This leads to the Chondritic Earth Model (Section 5.4.1).

(c) The meteorites derive from the asteroids by collision.

(d) The differentiated meteorites were formed within minor planets, or asteroids, which heated sufficiently to segregate into layers, forming an iron core, silicate mantle and transitional region between. Subsequent break-up due to collisions produced iron, achondrite and stony-iron meteorites.

4.6   Nucleosynthesis and stellar evolution

4.6.1   *Nucleosynthesis*

The starting point is assumed to be hydrogen, for it is both the simplest and by far the most abundant of the elements. To cause nuclei (see Note 7) to combine, they have to be brought very close together, and this is difficult because of the strong repulsion of their positive electric charges. It can be achieved by the random thermal velocities attained in a 'gas' heated to many millions of degrees. Since progressively heavier nuclei have larger electric charges, which increase the repulsive force, a series of **thermonuclear** reactions, at progressively higher temperatures, is needed to build up the elements, as summarised in Figure 4.12. (Atomic particle accelerators can achieve the same velocities without high temperature, and these permit the reactions to be studied in the laboratory).

The first step is **hydrogen-burning**, in which four protons, or hydrogen nuclei, are united via a number of possible reactions at 2×10^{7}°C or more, to form one α-particle, or helium nucleus. At higher temperatures (about 2×10^{8}°C), **helium-burning** causes α-particles to combine, but, because ^{8}Be is unstable, three α-particles have to unite to form ^{12}C, which can then absorb a further one to form ^{16}O. These reactions exhaust the helium, and the temperature must rise to about 5×10^{8}°C before **carbon-burning** and **oxygen-burning** can commence. When they do so, they produce ^{28}Si, which is particularly stable, plus some other nuclei.

To make two ^{28}Si nuclei combine directly would require a further large temperature rise, but before this could be reached other reactions occur. These reactions, in effect, break pieces at random off some of the nuclei present and add

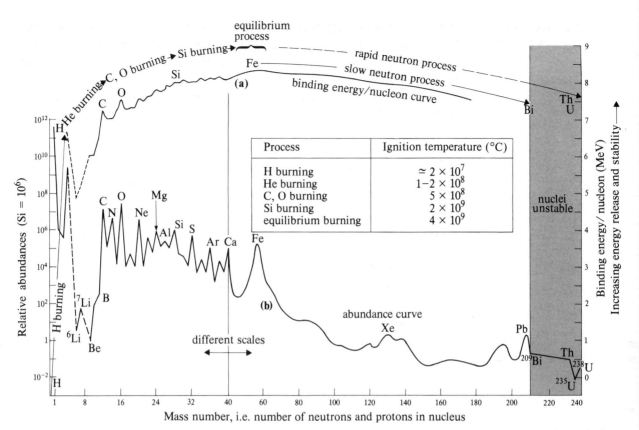

Figure 4.12 Nucleosynthesis and the abundance of the elements. (a) Binding energy per nucleon curve (upper curve): the higher up the curve that a nucleus is, the more stable it is, and, if a less stable nucleus is converted to a more stable one, energy is released, proportional to the vertical difference multiplied by the number of nucleons involved. The more important nuclear reactions are indicated. For further details see text and Note 7. (b) Relative abundances (lower curve): these are derived from solar and meteoritic abundances and are believed to represent the proportions in the Solar Nebula. Note that superimposed on the overall trend of decreasing abundances are peaks corresponding to peaks or changes of slope on the curve above. (Data from Cameron 1973.)

them to others, the net effect being to convert silicon to more stable and heavier nuclei. This is the **silicon-burning** process and, with a further rise of temperature, the reactions culminate in the equilibrium or **e-process**. The random e-process with its constant removal and addition of particles to and from the nuclei, is like shaking objects in a box: the shaking, provided it is not too violent, tends to rearrange the objects into a more stable arrangement. Similarly the e-process rearranges the nucleons to produce the most stable nuclei. These are iron and other elements at the summit of curve (a) of Figure 4.12.

It is not possible to build beyond the iron group of elements by increasing the temperature, because if any were produced they would be less stable and so soon be converted back to the iron group by the e-process. Instead, they can be produced by simple bombardment by neutrons which, having no electrical charge, are added easily to a nucleus. It is known from experiment that the neutron-to-proton ratio of a nucleus must remain within fairly narrow limits, which depends upon the size of the nucleus. In the slow-neutron or **s-process**, the neutrons are added at sufficiently long intervals so that, when a nucleus becomes unstable by having too large a neutron-to-proton ratio, it has time to change to a more stable form before a further neutron is added. This it does by internal conversion of a neutron to a proton with emission of an electron, or β-particle, to conserve electrical charge. This may take weeks or more, a very long time on nuclear time scales.

67

The s-process can build only as far as ^{209}Bi because the next heavier nucleus is unstable, no matter what neutron-to-proton ratio it has. This obstacle can be overcome by the **r-process** in which neutrons are added so rapidly that the nuclei formed do not have time to decay before a further neutron is added. In this way, much heavier nuclei can be built up. Of course, once the neutron bombardment has ceased, the neutron-rich nuclei will undergo repeated β-decays to produce nuclei that are relatively more stable but which, in turn, are unstable to α-decay and so break down to lighter nuclei. However, these will include ^{238}U, ^{235}U and ^{232}Th which have half-lives comparable to the age of the Earth and so have not yet decayed to negligible amounts. ^{244}Pu is an example of a heavy but shorter-lived isotope which was formed but has decayed away, and which was met in connection with meteorite formation intervals. The result of the r-process is to build past the highly unstable nuclei immediately above ^{209}Bi to the relatively more stable nuclei that lie beyond.

The s-process also helps fill in some of the gaps between lighter elements left by the other processes, such as between ^{12}C and ^{16}O. However, since it can produce only neutron-rich nuclei, a further process is needed to account for the known nuclei with lower than average neutron-to-proton ratios. This is the **p-process** that acts by adding protons.

Finally, how are the light elements Li, Be and B produced? Most of their isotopes are not formed in the above processes; in fact, they are destroyed at the temperature of hydrogen-burning. Probably they are formed as fragments when heavy nuclei in interstellar dust are struck by cosmic rays ('spallation products'). This has a very slow rate of production, but the nuclei spend a long time in space!

4.6.2 *Stellar evolution*

The most likely place for these nuclear reactions (except the last) to occur is in the interior of stars, because of the extremely high temperatures needed. In Section 4.3.2 we followed the evolution of an interstellar cloud into protostars, that is, fairly condensed masses at several thousand degrees. As explained, in this state, self-gravity can only be balanced by a high internal temperature, so that heat loss from the surface causes the mass to contract and heat up. Ultimately, the temperature rises high enough to initiate hydrogen-burning. This replenishes the heat lost from the surface, and the star no longer has to contract. Since nuclear hydrogen-burning is an extremely potent form of heat, the star enters into a prolonged stable period known as a 'main sequence' star, which in the case of the Sun is likely to last roughly 10^{10} years, about twice its present age. Yet, eventually, the hydrogen begins to be used up. This happens first at the centre, because the deep interior of a star does not convect and so cannot remove the accumulating helium. As a result, a core of helium is formed which grows in size, and the hydrogen-burning shell which surrounds it moves slowly outwards.

Since pressure is simply the combined effect of the impacts of all the particles in a 'gas', the reduction in number of particles by nuclear reactions, such as the conversion of four hydrogen nuclei into one helium nucleus, tends to lower the pressure inside a star. The star responds by contracting, but this results in a rise in temperature until helium-burning commences, leading in turn to a carbon–oxygen core, and so on. Just how far a star proceeds through the series of thermonuclear reactions depends upon its mass. A small mass of gas will never reach a sufficient temperature to burn its hydrogen, because, before it can contract sufficiently, the hydrogen atoms will come into contact with each other, so preventing further contraction and the mass will then cool. Such is roughly the case of Jupiter. The Sun, being much larger, can burn hydrogen, but not helium. But the largest stars, more than six times as massive as the Sun, will continue through the thermonuclear reactions, and then they consist of a series of concentric shells each burning heavier nuclei than the one outside it (Fig. 4.13).

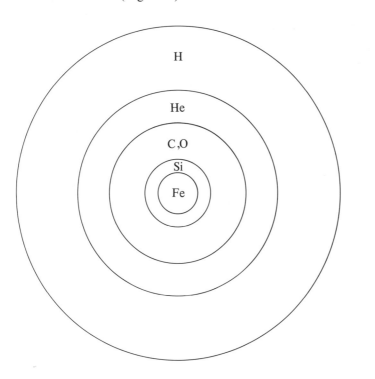

Figure 4.13 Section through an evolved massive star. This schematic figure shows how, with increasing temperature, nucleosynthesis has built up to iron at the centre, with earlier stages still proceeding in concentric shells. At this stage in its evolution, nuclear fuel is nearly exhausted at the centre and the star will soon become unstable and explode as a supernova.

Once the e-process has been completed at the centre of the star, with the production of the iron group of elements, there is no more energy available there from nuclear reactions (Fig. 4.12, curve (a)), though burning still continues in the surrounding shells. The iron-group core contracts, as with the earlier cores, causing a rise in temperature. The star remains stable until the temperature reaches about $10^{10}\,°C$, when the iron-group nuclei begin to break down into simpler particles, such as α-particles and neutrons. (This is analogous to shaking a box of objects so hard that they are shaken out.) This disintegration, of course, *absorbs* a great deal of energy, which the star attempts to provide by further contraction, but this merely aggravates the situation.

The pace of evolution of the star has been quickening all along. The curve of binding energy per nucleon (Fig. 4.12, curve (a)) shows that most heat is released by the conversion of hydrogen to helium, with subsequent stages releasing less and less heat. Coupled with the rising temperature and increasing surface heat loss, this results in each stage being passed through much more quickly than the preceding one, and it is estimated that the equilibrium stage may take only minutes. The final stage of core contraction, the breakdown of iron-group nuclei, becomes a collapse, and a star that has been burning nuclear fuels for millions of years may find itself bankrupt in seconds.

The collapse of the interior of the star leaves the outer parts unsupported, so they too contract and heat up rapidly. But they still contain much unburnt fuel and, since nuclear reactions are extremely sensitive to temperature, a runaway situation develops in which burning raises the temperature which, in turn, increases the rate of burning. The star has no time to adjust and much of the remaining fuel is consumed in a fraction of a second, blasting the outside of the star into space.

This catastrophic end to a star's evolution is thought to be one cause of a supernova in which, for a few weeks, it can be as bright as a whole galaxy of stars. The best-known supernova was observed by the Chinese in AD 1054, and the still-rapidly-expanding fragments are visible as the Crab Nebula (Fig. 4.5b). (The term 'nebula' is applied to many types of astronomical object. It just means a hazy patch of light ('nebulous'), which was how they appeared in the earlier telescopes.) At its centre is a pulsar, or neutron star, identified as the core collapsed to the density of a nucleus and only a few kilometres across.

How many of the naturally occurring isotopes can a supernova produce? Clearly, it can produce many of those up to and including the iron group before it reaches the catastrophic stage, but what of those beyond? Examination of the relative abundances of the isotopes between ^{56}Fe and ^{209}Bi shows that they were produced by the s-process, not the r-process. This can be deduced because the s-process, which allows nuclei with one-too-many neutrons time to decay, produces certain isotopes that the r-process cannot produce, since further neutrons are added before β-decay can occur. But building ^{209}Bi from the iron group takes several hundred years, because of the times required for the β-decays, and this time is not available in the final stages of a pre-supernova star. The solution to this problem seems to be that the s-process proceeds at the same time as some of the earlier reactions in the evolution of a pre-supernova star, and operates on the small amount of iron group isotopes incorporated in the star when it first formed from the interstellar cloud. The isotopes beyond ^{209}Bi are produced by the r-process in the last moments before the supernova explosion occurs and blasts them out in space, to join the interstellar gas and dust.

This explanation for the synthesis of the heavier elements only pushes the problem back a stage, for whence came the iron-group elements in the interstellar medium? The answer is from previous supernovae. This leads to, at least, a two-generation scheme for element production, hydrogen being built up to the iron group and then returned to the interstellar medium by one supernova, to be incorporated in another and processed further. In fact, it is believed that many supernovae were involved in the production of the elements of which the Earth and man are made.

Pre-supernova stars burn their fuel proportionately much faster than smaller stars, because of their much higher internal temperatures. Thus, between the time when the Galaxy was formed about 15×10^9 years ago and the time when the Solar System formed about 4.6×10^9 years ago, there was time for many generations of pre-supernova stars to form, explode and slowly enrich the interstellar medium, and hence later stars, in the heavier elements that form 2% of the Solar System.

A gradual build-up of the heavier elements does not account for the meteorite formation ages based on ^{244}Pu, ^{129}I and ^{26}Al (Section 4.5.4). One explanation of these is that an interstellar cloud, already enriched by many supernovae, entered a galactic spiral arm. The compression did not cause star formation in the cloud but produced a star nearby which rapidly evolved into a supernova, contributing a further quantity of elements. This was repeated at the next passage through a spiral arm (likely to have been 100 Ma later). The formation age of 100 Ma deduced from ^{244}Pu and ^{129}I reflects the first passage, while the much shorter one based on ^{26}Al reflects the second passage.

To produce the observed amounts of ^{26}Al and other isotope anomalies, and a formation age of only 2–3 Ma, the latter supernova must have been close in space and time to the cloud, and this suggests it triggered the formation of the Solar System. This is supported by the observation that new stars sometimes form on the expanding front of a supernova explosion, which causes a local compression.

In a nutshell, the heavier elements of the Solar System accumulated over vast lengths of time, but were topped up by two supernovae, one just before the Solar System formed.

A final topic is to what extent the composition of the outer, visible part of the Sun is changed by nuclear reactions within. This is believed to be very small because the nuclear-burning parts do not convect, as do the cooler outer parts. However, lithium is destroyed at only 3×10^7°C, less than the temperature of hydrogen burning, and it is probable that it is convected deep enough to be destroyed slowly. This would account for why Li is depleted in the Sun and other old stars, relative to young stars and to meteorites and the Earth's crust (Figs 4.7 and 5.4).

Of course, this account of nucleosynthesis theory has been considerably simplified. We have given little consideration to the critical role that a star's size plays in its evolution and have ignored, for example, the fact that some stars become supernovae even at the carbon–oxygen-burning stage. However, we have been able to understand (i) the relative abundances of elements in the Sun (Section 4.4), where nuclear burning has hardly changed the original composition apart from conversion of hydrogen to helium in the deep interior, (ii) why the major elements of the Earth and meteorites are likely to be oxygen, silicon, magnesium, sulphur and iron (see peaks on curve (b), Fig. 4.12), and (iii) how the formation of the Solar System may be related to a particular supernova explosion.

Summary

1   By looking at the Solar System as a whole, it is possible to pick out the major features with which to test possible theories of its formation. As a result, it is concluded, though not dogmatically, that it formed from a solar nebula, a hydrogen-rich rotating cloud of gas and dust. Although planetary systems cannot be the inevitable result of the development of such a nebula – if only because planetary systems cannot form about binary stars, which are in the majority – it is generally assumed that the Solar System was not formed as a result of exceptional circumstances, as required by the early tidal theories. However, its formation may be somewhat unusual in having been triggered by a supernova. The bulk of the Solar Nebula contracted to form the Sun, but the perimeter was

prevented from contracting by its large angular momentum and this formed into the planets.

2   The solar abundances of the elements are close to those of the Solar Nebula, apart from lithium. Very broadly, the compositions of the planets can be understood as deriving from solar composition by progressive loss of volatiles, Jupiter having lost little, while the terrestrial planets retain only the most refractory elements and compounds. This will be discussed further in the next chapter.

3   The chondritic meteorites – though possibly overstressed in default of other information – provide a valuable glimpse into the early stages of accretion. Their elemental abundances are close to those of the Sun, after allowing for loss of the most volatile elements and compounds, and detailed study of their composition and structure leads to the Chrondritic Earth Model (Section 5.4). The differentiated meteorites have undergone segregation processes, probably in asteroids, which may have been similar to those that took place in the Earth.

4   Consideration of stellar evolution and nucleosynthesis adds support to the above conclusions, and explains why oxygen, silicon, magnesium, sulphur and iron are probably the most abundant elements in the Earth.

Further reading

General journals:

  Scientific American (September 1975): description, theories of formation of the Solar System.

  Schramm & Clayton (1978): isotope anomalies in meteorites.
General books:

  Williams (1975): theories of formation of the Solar System.

  Wood (1979): the Solar System.

  Wood (1968); McCall (1973): meteorites.

  Tayler (1972a): nucleosynthesis.

  Tayler (1972b): stellar evolution.

  Mitton, S. (1977): *The Cambridge encyclopaedia of astronomy*. London: Jonathan Cape.
Advanced journals:

  Williams & Cremin (1968): systematic features and theories of formation of the Solar System.

  Huang (1973): systematic features of the Solar System.

  Reeves (1974): conference report on the Solar System: lively insight into how the subject advances.

  Icarus **42**, 29–34 (1980): Pluto.

  New Scientist **79**, 273 (1978): Pluto.

  Trimble (1975): abundances of the elements, and nucleosynthesis.
Advanced books:

  Dermott (1978): theories of formation of the Solar System.

  Wasson (1974); Nagy (1975): meteorites.

  Shklovskii (1978): stellar evolution.

5 The accretion and chemical layering of terrestrial planets

5.1 Introduction

In Chapter 4, we were concerned with the origin of the chemical elements and the physical conditions pertaining to the pre-Solar Nebula. We began to consider the formation of planetisimals due to the condensation of solid grains from vapour, and their aggregation by magnetic, by electrostatic and, most important during the later stages, by gravitational attraction into substantial planets. The composition of the modern solar atmosphere was compared with that of chondritic meteorites in Figure 4.7. The close similarity in the abundances of elements heavier than oxygen leads to the suggestion that some of the resulting planets might also have chondritic bulk (i.e. total average) compositions. The Chondritic Earth Model proposes that this is true for our own planet – a proposition that will be evaluated rigorously in Section 5.4.1.

The next stage is to examine in more detail the accretion of planets, because this leads to certain predictions about chemical variations that may occur between and inside planets. There are two contrasting views of the origin of planetary chemistry, namely those of **homogeneous** and **heterogeneous** accretion. These two processes can be evaluated by testing their predictions of planetary chemistry against the end-products of accretion – the planets themselves. Unfortunately, our supply of data on planetary compositions is scant. Apart from quite detailed information about the Earth, we have surface sample analyses only from Mars and the Moon. Our knowledge of the chemistry of the other planets is estimated indirectly from mass, volume and moment of inertia and from general ideas about the formation of the Solar System.

In spite of these difficulties, it is known that most of the terrestrial planets probably do have rather similar chemistries and that, apart from Mercury, they may not differ substantially from the bulk composition of undifferentiated chondritic meteorites. But was accretion on a planetary scale homogeneous, with the *subsequent* development of layers or has layering *resulted directly* from heterogeneous accretion processes? Certainly, there is good evidence within the Solar System, from the outer planets, their satellites and Mercury, that planetary compositions outside the terrestrial planet group vary quite markedly and that fundamental heterogeneity in the primitive Solar Nebula must have preceded planet formation. But, in debating the origin of the Earth, we are more concerned about accretion on the scale of the Earth/Venus/Mars group and the individual planets themselves: Was this a homo-geneous or a heterogeneous process? This question is considered in the following pages, first, by discussing the causes and consequences of element segregation (i) before planetary accretion and (ii) during or after accretion (Sections 5.2 and 5.3), and, secondly, by testing predictions based on (i) and (ii) against planetary data (Sections 5.4 and 5.5).

5.2 Models of planetary accretion and layering

5.2.1 *Pre-accretion element segregation: the heterogeneous model*

There can be little doubt that some chemical separation resulted from selective condensation of different elements and compounds from the high-temperature, low-density primitive Solar Nebula gas cloud. For example, the relatively low densities and large sizes of the outer planets (Table 4.1) indicate that these planets contain a much larger proportion of the relatively volatile light elements such as H, He, C and O than do the terrestrial planets. This is to be expected if there was a radial decrease in temperature away from the Sun (Section 4.3.2). Another example occurs in the Allende C3 chondritic meteorite (Section 4.5.4). The small white inclusions contain a high proportion of calcium- and aluminium-rich materials known to condense first from a low-density gas of nebula composition. This demonstrates that the Allende material must have gone through a totally gaseous stage before its source meteorite accreted, and the same may well be true of the entire primitive Solar Nebula. In contrast to the normal temperatures of interstellar space (from about -150 to $-200°C$), temperatures of at least $1250°C$ were reached, mainly by gravitational heating, before planet formation.

Grossman and Larimer (1974) determined experimental condensation temperatures for the Solar Nebula, given the relative vacuum (10^{-4} atm) that probably existed during condensation.

These are recorded in Table 5.1 as the likely mineral constituents that control the ultimate chemistry of the condensates. This table is highly simplified, and it should be noted that different minor elements can substitute in these mineral structures causing small changes in the appropriate temperatures. **Refractory** materials are those which condense above $1000°C$ and these include the Allende-type Ca- and Al-

Table 5.1 Approximate condensation temperatures of Solar Nebula materials at 10^{-4}atm. (After Grossman & Larimer 1974.)

Mineral phase	Composition	Condensation temperature (°C)
corundum	Al_2O_3	1410
melilite	$Ca_2Al_2SiO_7-Ca_2MgSi_2O_7$	1205
perovskite	$CaTiO_3$	1200
spinel	$MgAl_2O_4$	1150
metallic iron	$Fe(Ni)$	1130
forsterite	$Mg_2SiO_4 (-Fe_2SiO_4)$	1120
diopside	$CaMgSi_2O_6$	1100
enstatite	$MgSiO_3 (-FeSiO_3)$	
anorthite	$CaAl_2Si_2O_8$	
		refractory
		volatile
alkali feldspar	$(Na,K)AlSi_3O_8$	980
troilite	FeS	430
magnetite	Fe_3O_4	135
ice, methane, etc.	$H_2O, CH_4, CO_2, O_2, N_2, H_2$, etc.	<0

rich condensates, forming at the highest temperatures, followed by metallic iron and iron-magnesium silicates. Materials still **volatile** below 1000°C include the alkali elements (although these may be incorporated into feldspar that has already formed), iron sulphide and iron oxide. At the lowest temperatures, materials we normally regard as gaseous at the Earth's surface (about 0–30°C) condense from low-density. gas.

Table 5.1 indicates that prior to or during accretion most meteoritic and terrestrial planet material must have condensed in the temperature range down to about 100°C, whereas the material in the outer planets would not have condensed until the temperature had dropped well below 0°C. This brings us back to the radial temperature gradient mentioned earlier. But the proponents of heterogeneous planetary accretion claim also that the temperature fell *locally* and that condensates were forming all the time that planets were accreting. This means that, at any given distance from the Sun, materials segregated progressively into solids in order of their condensation, resulting in the present layered planetary structures. In the lower-density outer planets, silicate and iron cores developed and preceded the condensation and accretion of large gaseous envelopes. In the rocky, high-density terrestrial planets, the early stages are magnified: according to heterogeneous accretion, a high-temperature core rich in calcium and aluminium formed first, followed by metallic iron and, finally, the mantle silicates, forsterite, diopside, anorthite, etc. (Table 5.1). Thus, a three-layered heterogeneous stage is likely, as shown in Figure 5.1a. Later on, it is envisaged (as in the homogeneous model, Section 5.2.2) that as the interior warms

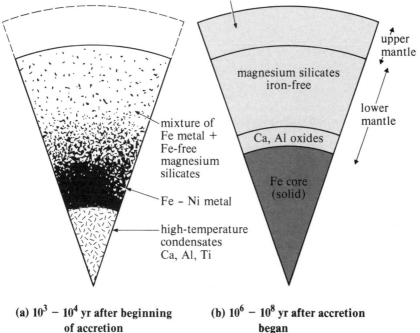

mixture of Mg silicates + Fe_3O_4 + hydrated silicates + other low-temperature condensates; depleted in Ca, Al

upper mantle

magnesium silicates iron-free

lower mantle

Ca, Al oxides

Fe core (solid)

mixture of Fe metal + Fe-free magnesium silicates

Fe – Ni metal

high-temperature condensates Ca, Al, Ti

(a) $10^3 - 10^4$ yr after beginning of accretion

(b) $10^6 - 10^8$ yr after accretion began

Figure 5.1 Heterogeneous accretion of a layered terrestrial planet (a) in the late stages of accretion showing element zonation according to condensation temperatures (Table 5.1) and (b) just after accretion as internal element sorting by density differences accompanies early heating of the planet. Process (b) must lead ultimately to the zones shown in Figure 1.1. (After Clark *et al.* 1972. From *The nature of the solid Earth*, E.C. Robertson (ed.) © McGraw-Hill Book Company.)

75

up sufficiently for melting and element segregation according to density to occur, a dense, central, iron-rich core forms as the lighter, high-temperature condensates are displaced back into the silicate exterior (Fig. 5.1b). The difficulty with this part of the heterogeneous model is that the details of multi-layer accretion cannot be tested by planetary observation because the layers are transient features. Good scientific theories should make predictions that can be tested, but this aspect of the heterogeneous accretion model is uncertain and indeterminate.

However, radial temperature gradients in the nebula are testable, and it was noted earlier that the outer planets finished condensing at lower temperatures than the terrestrial planets. But, it also follows that there should be variations in composition and, therefore, density *within* the terrestrial planet group. If pre-accretion element segregation was significant on this scale, then planets further from the Sun should contain larger proportions of alkali feldspars, troilite, magnetite, etc., which condense at low temperatures. In addition to increases in alkali elements in this direction, there should also be *decreasing iron/silicon ratios* and hence, *mean density* (since iron is very much more dense than silicates). We shall examine these predictions in more detail, together with planetary data, in Section 5.3; in the meantime, the model is summarised in the middle column of Table 5.2.

5.2.2 *Homogeneous planetary accretion with subsequent layering*

In this model, it is envisaged that all the materials that accreted to form each planet had already condensed into grains and, in the case of the rocky terrestrial planets, were at some temperature below 100°C (Table 5.1). It is also assumed that these grains were thoroughly mixed together and, in the case of the Chondritic Earth Model (Section 5.4.1), had a composition not unlike C1 chondrites. There is no reason to suppose that any changes occurred in the supply of chemical constituents as the grains accreted. However, there are some important mechanisms which may have segregated elements subsequently due to *reheating* both during and after accretion.

Initially, there is reheating due to accretion itself, because the gradual falling-in of grains to form planets results in a loss of kinetic energy, much of which must appear as heat. In the early stages of accretion, gravity is low and impact velocities for newly accreted grains are small, but, as a protoplanet grows, impact velocities increase and the temperature rise becomes much more significant. Some of the condensed material may be revolatilised; in particular, oxygen, sulphur, carbon and the alkali elements may be released, principally from silicates, at the accreting surface and lost into space as various oxide combinations (H_2O, CO_2, SO_2, etc.). The loss of such oxygen-rich compounds by volatilisation must be accompanied by chemical reduction in the remaining silicate material. For example, in an Fe–Mg olivine mixture, *without changing the Fe/Si ratio in the mixture*, the amount of iron in silicate can be reduced and the amount of free iron increased by removing oxygen as follows:

$$Mg_2SiO_4 + Fe_2SiO_4 \longrightarrow 2MgSiO_3 + 2FeO \qquad (5.1)$$
$$\text{olivine} \qquad\qquad \text{pyroxene}$$

accompanied by

$$2FeO + C \longrightarrow 2Fe + CO_2\uparrow \qquad (5.2)$$

Table 5.2 Properties of terrestrial planetary accretion models.

Features	Heterogeneous accretion	Homogeneous accretion
accretion temperatures	incandescent, above highest condensation temperatures and falling during accretion; incandescence due to heating during contraction of Solar Nebula under gravity	relatively cold, within the volatile condensation range of Table 5.1
cause of chemical difference between planets	selective condensation during nebular cooling: hence heterogeneous; lower-temperature condensates are favoured by greater distances from nebula centre	volatilisation during and after accretion due to planetary heating; accreting material for all planets is homogeneous; amount of volatilisation increases with size of planet
reduction of iron to form core material	condenses directly as metal at temperatures greater than for silicates and sulphides of iron	reduction of silicates, etc., effected near the accreting planetary surface during initial heating and volatilisation (Eqs 5.1 & 5.2)
timing of element segregation	pre-accretion followed by readjustments due to post-accretion melting and longer-term changes	post-accretion due to initial heating and melting followed by longer-term changes
cause of internal planetary heating	initially hot and cooling, but also reheated as in homogeneous model	release of kinetic energy during accretion, early short-lived radiogenic heat and long-term, long-lived radiogenic heat
planetary layering	selective condensation into a layered structure due to temperature gradient across the nebula and possible falling temperature during accretion	iron-rich core material becomes molten near the surface due to initial heating and sinks, leaving a solid silicate mantle
predicted chemical/density variations between planets, assuming chondritic starting materials	planets have progressively lower Fe/Si and refractory/volatile element ratios further away from the Sun; density therefore decreases in this direction	planets have similar total Fe/Si ratios but vary in volatiles and oxidation as a function of planetary size; small planets retain more volatiles, are more oxidised and have lower densities than larger ones

Both volatile loss and reduction processes (Eqs (5.1) and (5.2)) become increasingly effective with growth of protoplanet size. Starting from the same material the most massive planets should have the least-oxidised compositions, having been most effectively heated by accretion. Because oxygen is one of the lightest elements in terrestrial planets, this model makes the specific prediction that *the largest planets should have the highest densities* and should contain the lowest concentrations of the relatively volatile alkali elements. It is very important that these predictions differ from those of the heterogeneous accretion model (Section 5.2.1) in which distance from the Sun rather than size was critical.

A second mechanism for planetary heating, once accretion has occurred, is due to radioactive decay. Evidence was presented in Section 4.5.4 that relatively short-lived radioactive isotopes such as ^{244}Pu, ^{129}I and ^{26}Al probably were present in the primitive Solar Nebula, and that significant quantities of these isotopes became incorporated into meteorites. Timing is a major uncertainty and yet is critically important. For example, if the whole Earth had formed as early as the Allende high-temperature condensates, it would have contained about 0·5 parts per million of ^{26}Al – enough to melt the entire planet (see O'Nions *et al.* 1978)! But a delay of 0·7 Ma (the half-life of ^{26}Al) would have halved the rate of heat release.

It is now clear that two processes could have heated the early planets: accretion and the subsequent decay of short-lived radioisotopes. As heating continued, perhaps backed up by radioactive decays following accretion, melting took place. The reduced iron-rich materials (Eqs 5.1 and 5.2) are first to melt, between 1000 and 1500°C, while silicates are still solid. Because of chilling at the planetary surface due to radiation of heat into space, a rigid insulating crust is thought to have formed around each planet soon after accretion. Therefore, the molten material must have collected at some depth below the surface, perhaps as pools or drops – see Figure 5.2. As the drops grew more massive the size of the gravitational force upon them became more effective, and they fell towards the centre of the planet, permeating through a mass of silicates. The sinking process released more gravitational energy as heat, initiating a runaway process that resulted in a molten core, predominantly composed of iron, and left a solid silicate mantle (see Chapter 6 for further details of melting conditions, etc.). Thus, initially homogeneous terrestrial planets may have become internally layered into the major divisions of core and mantle, early in their history (Table 5.2).

Figure 5.2 Successive stages of drop formation from a heavy liquid layer, used here as a model for core formation in an initially homogeneous planet. (After Elsasser 1963.)

(a)   **(b)**

All these events took place soon after the Solar System formed, 4600 Ma ago. Since then, the planets have undergone further partial melting and element segregation processes, partly resulting from their initial heating and partly because of long-term radiogenic heating through ^{238}U, ^{235}U, ^{232}Th and ^{40}K decays. In particular, the growth of a chemically distinct crustal layer is known to have taken place on the Earth and the Moon, and is postulated for Mars and Venus. Unlike the rapid growth of planetary cores, there is good evidence that crustal growth took a substantial part of each planet's history (see Chs 7–10 for the Earth). However, the duration of this activity appears to be a function of planetary size (see Section 5.4) and is related to the ratio of volume to surface area, which determines the ratio of heat content to radiative heat loss (Kaula 1975).

The main features of the heterogeneous and homogeneous accretion models are summarised in Table 5.2. We emphasise that these are two extreme viewpoints: there have been many other feasible proposals which combine the two approaches in different ways (see Section 5.5 for a possible combination).

A feature which is common to nearly all models is the heating that occurred soon after accretion due to a variety of possible causes. This is necessary to separate the core-forming elements in the homogeneous model and to sort them in the heterogeneous model (see Fig. 5.1), both according to density. But the chemical and density differences that would be anticipated using the latter model are much more fundamental and far-reaching due to the high-temperature accretion process. The heterogeneous model proposes that iron is accreted directly as reduced metal, whereas the homogeneous model starts with the most oxidised material which was identified (Section 4.5) as the most primitive and representative of the Solar Nebula: carbonaceous chondrite meteorites. Referring to Figure 4.9 p. 61, we find that there are major variations in the oxidation state of meteorites from the least oxidised (E group), which contain virtually no iron silicates, to the most oxidised (C group), from which metallic iron is absent. More important there are quite distinct Fe/Si ratios for most groups, increasing in the order LL, L, C, E and H (the latter two are indistinguishable). It would seem that different condensation temperatures are needed – as in the heterogeneous model. However, there is good reason to suppose that most meteorites may not be equilibrium assemblages, because they had relatively short histories at high temperature (Section 4.5). Although the variation of Fe/Si ratios could indicate heterogeneous accretion, it is equally possible that the least-oxidised meteorites (E and H) also contain the highest Fe/Si ratios *because* they were frozen part-way towards core formation in their parent planets. To be more decisive about choosing between these accretion models, it is necessary to move from small, and possibly unrepresentative, meteorite samples to the massive terrestrial planets themselves, including the Earth and the Moon. But, before we do that, it is useful to establish a few geochemical pointers to the relative volumes of silicate, sulphide and metallic zones within the planets, and the distribution among them of all the chemical elements.

5.3 Element segregation: some geochemical rules

Many geological processes, such as weathering, sedimentation, metamorphism and partial melting, continually separate the chemical elements in what has been termed 'the rock cycle'. For instance, segregation among sediments by physical and chemical processes leads to sandy beaches (SiO_2), limestone reefs ($CaCO_3$) and coal measures (dominantly carbon). The way in which elements behave both at the surface and within planets is controlled largely by their electronic configurations and their affinities for different types of crystalline bonds. In differentiated meteorites, for example, the elements are sorted into clear, though overlapping, groups based on silicate, sulphide and metal, and these are distinguished as follows:

(a) *Lithophilic* elements, which tend to occur with oxygen in oxides and silicates, e.g. Rb, K, Ba, Na, Sr, Ca and Mg [Gr. *lithos*, stone].

(b) *Chalcophilic* elements, which tend to concentrate as sulphides, e.g. Cu, Zn, Pb, Sn and Ag [Gr. *khalkos*, copper].

(c) *Siderophilic* elements, which tend to be metallic, e.g. Fe, Ni, As, Pt, Ir and Au [Gr. *sideros*, iron].

The same rules are applicable to all terrestrial rocks and, clearly, they are related to fundamental properties of the elements. In fact, the bond-forming ability of elements is a function of their position in the Periodic Table. Here we will simplify matters by using just one property of each element, its **electronegativity**, E, measured on a dimensionless scale from 0 to 4 devised by Pauling (1959). This is the ability of an atom to attract electrons and so to become a negatively charged **anion**. Thus, the value is highest for halogens (F^-, 4·0), it is also high for oxygen (O^{2-}, 3·5), and is moderate for sulphur (S^{2-}, 2·5). But it is low for the more metallic elements which lose electrons to form positively charged **cations** (e.g. Mg^{2+}, 1·2; Si^{4+}, 1·8; Fe^{2+}, 1·8). Apart from oxygen, sulphur and certain complex anions (e.g. silicates, phosphates, etc.), all the elements in which we are interested form cations, and their electronegativities are given in Table 5.3.

Table 5.3 Geochemical affinities and cation electronegativities. (After Pauling 1959.)

$E < 1·6$ Lithophilic		$1·6 < E < 2·0$ Chalcophilic		$2·0 < E < 2·4$ Siderophilic	
Cs^+	0·7	Zn^{2+}	1·6	As^{3+}	2·0
Rb^+	0·8	(U^{4+}	1·7)	(P^{5+}	2·1)
K^+	0·8	(W^{4+}	1·7)	Ru^{4+}	2·2
Ba^{2+}	0·9	(Si^{4+}	1·8)	Rh^{3+}	2·2
Na^+	0·9	(Ge^{4+}	1·8)	Pd^{2+}	2·2
Sr^{2+}	1·0	Fe^{2+}	1·8	Os^{4+}	2·2
Ca^{2+}	1·0	Co^{2+}	1·8	Ir^{4+}	2·2
Li^+	1·0	Ni^{2+}	1·8	Pt^{2+}	2·2
rare earths	1·0–1·2	Pb^{2+}	1·8	Au^+	2·4
Mg^{2+}	1·2	Mo^{4+}	1·8		
Sc^{3+}	1·3	Cu^{2+}	1·9		
Th^{4+}	1·3	Ag^+	1·9		
V^{3+}	1·4	Sn^{4+}	1·9		
Zr^{4+}	1·4	Hg^{3+}	1·9		
Mn^{2+}	1·5	Sb^{3+}	1·9		
Be^{2+}	1·5	Bi^{3+}	1·9		
Al^{3+}	1·5	Re^{3+}	1·9		
Ti^{4+}	1·5				
Cr^{3+}	1·6				

Bracketed elements are those that tend to be lithophilic because their small ionic size and large charge favours the formation of complex anions with oxygen – hence they are misclassified using simple element electronegativities.

If there is an appreciable difference in the values of E for two elements, then a bond formed between them is likely to be **ionic**, characterised by strong electrostatic attractions (NaCl, for example). All the elements in Table 5.3 with values of E less than 1·6 have an affinity for ionic bonding with oxygen and so exhibit lithophilic behaviour in terrestrial planets. Apart from the elements listed in parentheses in Table 5.3, which are also lithophilic because their large charge promotes the formation of complex anions with oxygen, the elements that have E values falling between 1·6 and 2·0 tend to be chalcophilic. This is because small differences in electronegativity, such as between these elements and sulphur ($E = 2·5$), favour the 'sharing' of electrons and hence **covalent** bonding. In fact, there is a complete range of bond types from completely ionic to completely covalent, determined by variations in the electronegativities of combining elements. The third type of naturally occurring bond is that which occurs in **metals** which have closely packed regular structures, often containing just one type of atom. The positively charged metal ions are surrounded by a 'gas' of mobile electrons, which gives rise to the high thermal and electrical conductivities of metals. Elements whose electronegativities lie between 2·0 and 2·4 belong to this group and are termed siderophilic because they tend to occur along with metallic iron in the Earth. This raises an interesting paradox because iron, which gives its name to the siderophilic group, has an electronegativity of 1·8 and so should be a chalcophilic element according to these definitions. Before explaining this paradox, here is a summary of the relationship between electronegativity and geochemical affinities:

Lithophilic: $E < 1·6$ but, in addition, anion complex-forming elements of higher E value
Chalcophilic: $1·6 < E < 2·0$, except for those forming complexes
Siderophilic: $2·0 < E < 2·4$

(In terms of the Periodic Table, the most strongly lithophilic elements are cations from the alkaline and alkaline-earth groups, whereas chalcophilic and siderophilic elements belong mainly to the transition groups, with siderophilic tendencies also favoured by high mass numbers.)

What can be deduced from these geochemical rules about the distribution of elements inside the terrestrial planets? This depends on the relative proportions of the major elements present; in a planet with the composition of carbonaceous chondrites (shown in Fig. 5.3) the five elements, iron, oxygen, silicon, magnesium and sulphur, alone comprise about 92% by weight. If the elements are allowed to form compounds, then a lithophilic layer will develop dominated by magnesium silicates (olivines and pyroxenes). Referring to the relative masses in Figure 5.3, it might seem that there is not enough oxygen to combine with all the magnesium and silicon, but this is not so, because oxygen is a light element. Therefore, some will remain when all the magnesium and silicon have been used up. Because of its electro-negativity, iron would prefer to be chalcophilic and so it will automatically use up the available sulphur. But some iron is forced to combine with the remaining oxygen and thus to be lithophilic. However, because of the major abundance of iron (Fig. 5.3), the chances are that some iron will remain as metal after all the oxygen and sulphur have been used up, thus forming a siderophilic layer in the terrestrial planets. This is especially true if oxygen has been lost by volatilisation (Eqs (5.1) and 5.2)), and so *the oxygen content of a planet will determine the size of the siderophilic layer.*

81

Figure 5.3 Approximate concentrations (wt%) of the major chemical constituents in the solid materials of a carbonaceous chondrite planet. Organic compounds, water and other elements comprise the remaining 8%. (After Wiik 1956.) *N.B.* The abundances indicated here are those derived in Ch. 4 from studies of numbers of atoms in meteorites, except that here they are converted to relative atomic masses by dividing by atomic weights. This increases the apparent abundance of iron but otherwise does not affect the order of element abundances deduced in Ch. 4.

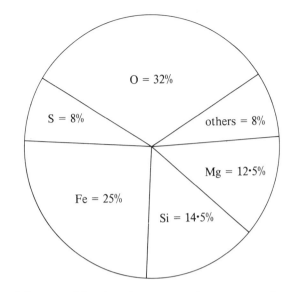

The result of these combinations is to create three separate layers dominated, in turn, by Mg–Fe silicates, FeS and Fe metal. Provided the temperature is high enough, these layers will separate in order of density. Thus the Earth's core is thought to comprise a metallic inner region, a sulphide-rich outer core and a silicate crust-plus-mantle (see Fig. 1.1 and Chs 6–8). These are the siderophilic, chalcophilic and lithophilic layers and the remaining metallic elements (M) will become partitioned into these layers according to their electrochemical properties (Table 5.3). Equilibrium reactions such as

$$\text{M sulphide} + \text{Fe silicate} \rightleftharpoons \text{M silicate} + \text{Fe sulphide} \qquad (5.3)$$

and

$$\text{M} + \text{Fe sulphide} \rightleftharpoons \text{M sulphide} + \text{Fe} \qquad (5.4)$$

will have operated throughout the active history of the terrestrial planets, continuously sorting out the three categories of elements into their most compatible layers. In the Earth, the efficiency of this process has been high, favoured by a long history of internal activity, but separation is still not complete. For example, traces of gold and platinum are found in the Earth's crust even though these elements must be strongly concentrated into planetary cores – an intriguing prospect for the ingenious entrepreneur? On a more serious note, one can predict from Table 5.3 that, in any planet, elements such as aluminium, calcium and the alkalis should be confined almost entirely to the mantle and crust, whereas nickel should enter the core.

Finally, it is worth reiterating that the division into layers depends critically upon the amounts of available oxygen and sulphur. If the Earth were made of literally carbonaceous chondrite material (as in Fig. 5.3), which is strongly oxidised, it would be almost entirely lithophilic with a small chalcophilic core (about 20% of its mass as FeS) and no free metal. Because the Earth has quite a large core (32% of its mass), it follows that, if in all other respects its chemistry is akin to carbonaceous chondrites, it must have lost a few percent of oxygen. Thus, the overall oxidation state of a planet may determine the size of its core, whether there is a metallic region, and the position of the boundaries between the respective layers.

5.4 The terrestrial planets and the Moon

5.4.1 *The Earth and the Chondritic Earth Model*

Here we look again at the Earth and the other nearby bodies in the Solar System, specifically to study those properties that usefully constrain accretion models. In particular, predictions about planetary compositions and densities in relation to size and/or distance from the Sun (lowest entry in Table 5.2) are of interest. As the Earth and the Moon have received most investigation it is useful to start with them and then to make comparisons with the other planets.

The Earth is the most dense and the largest terrestrial planet (Table 4.1) and this apparently accords with the predictions of the homogeneous accretion model. To test this conclusion further, we need to compare the bulk composition of the Earth with that of the primitive Solar Nebula (as for chondrites in Fig. 4.7). But this is not possible because our knowledge of its detailed composition is confined to direct observations of the crust, which is not typical of the whole Earth. However, using geochemical theory (Section 5.3), the differences between the composition of the well studied continental crust (Ch. 9) and the pre-Solar Nebula can be assessed and a worthwhile comparison made. This is shown in Figure 5.4 where the agreement between the compositions of the Earth's crust and the solar atmosphere is seen to be worse than for chondrites. The Earth's crust is derived by partial melting from the lithophilic mantle; these melts are preferentially enriched in lithophilic elements (Chs 7–9) and so the crust is the most strongly lithophilic region of the Earth. It is not surprising, therefore, that the lithophilic elements (Na, Al, Ca, K, Sr, Rb, Zr, Ba, etc.) are all strongly enriched in the crust, whereas chalcophilic (e.g. Zn, Cu, Cd, Ag) and

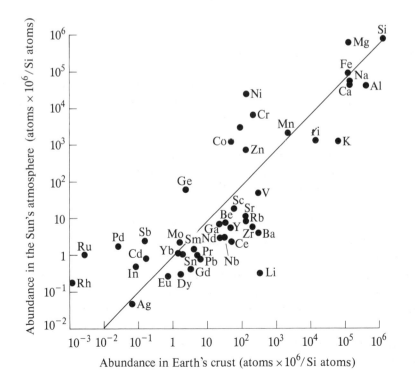

Figure 5.4 Comparison between element abundances in the solar atmosphere and in the Earth's continental crust, standardised as in Figure 4.7 by setting silicon at 10^6 on each axis. H, He, C, O and N are not plotted as they are far more abundant in the solar atmosphere but were either lost from, or only partially accreted to, the Earth. (After Wood 1968.)

siderophilic (Ni, Pd, Pt, Rh and Au) elements are depleted in the crust. Note that iron plots on the equal abundance line in Figure 5.4 and magnesium is the only lithophilic element to be depleted in the Earth's crust. The details of *mantle* melting processes (Chs 7 and 9) are such that iron concentrates into the first-formed partial melts, whereas magnesium remains in the residue (depleted peridotite, see Ch. 7). The iron/magnesium ratio, therefore, is much lower in the mantle than in the crust, and the lithophilic zone (crust plus mantle) of the Earth should be enriched in magnesium and depleted in iron relative to the solar atmosphere.

Bearing in mind that the siderophilic/chalcophilic core comprises 32% of the Earth's mass, and that the lithophilic crust and mantle are the remaining 68% (based on the density–depth discussion in Chapter 3), it follows from Figure 5.4 that the bulk composition of the Earth must be closely similar to the *non-volatile* components of the solar atmosphere and hence, via Figure 4.7, to chondrites. This was the basis for the Chondritic Earth Model, first proposed during the 1950s, which states that the bulk composition of the Earth is the same as carbonaceous chondrites (Fig. 5.3), given that most of the water and organic compounds have been lost. In view of the importance of such a model, which implies homogeneous accretion (Table 5.2), considerable attention has been focused on its validity and implications. Many complications have arisen, and it should be stated that the model must be regarded only as a good *approximation* to the composition of the Earth.

In 1963, Taylor decided to compare the crustal abundances of the elements with the composition of high-iron (H group) meteorites, because there are abundant data to show that they are like the Earth in that (i) these chondrites are relatively depleted of water and organic compounds compared with the carbonaceous group, and (ii) they contain most of their iron in the metal and sulphide forms. Taylor calculated the fraction of the total abundance of each element in an H-chondritic Earth which the *observed* crustal abundance would represent. Most strongly lithophilic elements reached significant fractions of their total anticipated terrestrial abundance within the crust alone, but in two cases, uranium and thorium, the crust actually contains more of the element (1·7 times and 2·1 times, respectively) than a chondritic Earth can supply! Another significant finding concerns the alkali elements, potassium, rubidium and caesium, which, because of their very strongly lithophilic tendencies, should reach between 50 and 100% of their terrestrial abundance in the crust. They should be concentrated at least as strongly as, for example, barium, lanthanum and cerium, which are slightly more electronegative (see Table 5.3). Proportions of chondritic abundances in the crust for K, Rb and Cs are 21, 26 and 22%, respectively, compared with Ba, La and Ce at 93, 82 and 59%. The conclusion is that K, Rb and Cs are depleted in the Earth to about one-third of their chondritic abundance. To summarise, this analysis suggests that the Earth contains:

(a) twice the U and Th of H chondrites, and
(b) one-third of the K, Rb and Cs of H chondrites.

A more recent review of the Chondritic Earth Model by Ringwood (1975) was based on the projected composition of the *mantle* rather than the crust, using evidence from ultrabasic rocks, basalt magmas, etc. (see Ch. 7, also O'Nions *et al.*

1978). Since the mantle comprises 68% of the Earth's mass, compared with $< 1\%$ for the crust, a comparison between chondrites and a mantle composition seems to be more valid. Ringwood obtained similar conclusions to those given above, but noted that the H group and most ordinary chondrites are depleted in refractory elements (Ca, Al etc.; Table 5.1), including U and Th, compared with carbonaceous chondrites. He also realised, with many other contemporary scientists, that Taylor had overestimated the U and Th contents of the crust. Therefore comparison of the bulk Earth with *carbonaceous* chondrites gives a much better match for the refractory elements, but still the alkali depletion discrepancy remains. Although there is some evidence that the 'missing' alkali elements could have been incorporated in the Earth's core (because of possible chalcophilic behaviour at high pressure (Ch. 6)), an Earth depleted of alkali elements would not constitute a difficulty for the homogeneous accretion model (Table 5.2). This is because alkali elements would volatilise during accretion after water, carbon dioxide, etc. Equally, the evidence from the Earth alone cannot disprove heterogeneous accretion because, according to the criteria of Table 5.2, we need to compare the composition of other planets with that of the Earth to make any positive statements in respect of variable Fe/Si ratios, etc. and this is the purpose of the following subsections.

5.4.2 *The Moon*

For centuries, the Moon has captured the imagination of scientists, but little information was available until the recent space exploration, which has resulted in new and exciting discoveries. Many of these concern the history of lunar activity in a geological sense and, more importantly, the composition of its interior. But, for these data to be most useful, it is necessary to consider the origin of the Moon.

The Moon has a radius of 1738 km and a mean density of $3340 \, \text{kg m}^{-3}$. It is, therefore, one of the largest satellites in the Solar System, and this has prompted the suggestion that it may once have been a planet in its own right. Historical records and geological evidence concerning the history of the Moon's orbit show that it is gaining angular momentum, hence distance, from the Earth and this indicates a closer association in the past. Opinions vary about whether the Moon accreted together with the Earth in space, or whether it fissioned from the Earth, or even whether it was an independent 'wanderer' in the Solar System, subsequently captured by the Earth (for reviews, see Kaula 1977, Wood 1978, J. V. Smith 1979). The binary co-accretion theory has proved most popular, mainly on dynamical grounds; for example, it is easier for the Earth to capture many small fragments than one large ready-formed Moon. It is instructive to consider the evidence for chemical differences between the Earth and the Moon against this background.

It is obvious from the Earth that the surface of the Moon has light and dark areas which are identified with topographic features: the lighter *highland* areas and the darker *mare* basins (so-called because they were once thought to be oceans). The highland areas comprise anorthosites and anorthositic gabbros (70–100% calcium-rich plagioclase) which are 4000 to 4500 Ma old. The Moon seems to have developed an early crust of this material which was disrupted when the mare basins were formed, probably by the impact of large planetisimals, 4000 Ma ago. The basins were subsequently filled with vast basalt lava flows, dating from

85

3900–3200 Ma old, but since that time no further volcanic activity is recorded.

The basalt lavas provide some of the most useful clues about the composition of the Moon. This is because, like terrestrial basalts, they must have been produced by partial melting of the lunar interior (see also Ch. 7). In Figure 5.5, the ratio of abundance in lunar to terrestrial basalts is plotted for certain elements against their condensation temperature (Table 5.1). The alkali elements are included as volatiles in Figure 5.5 and they are relatively depleted in lunar basalts, whereas refractory elements are enriched. Bearing in mind that nickel and iridium are two refractory elements that normally show siderophilic tendencies, then the best description of Figure 5.5 is as follows:

In comparison with terrestrial basalts, volatile and siderophilic elements are depleted, whereas refractory lithophilic elements are enriched in lunar basalts.

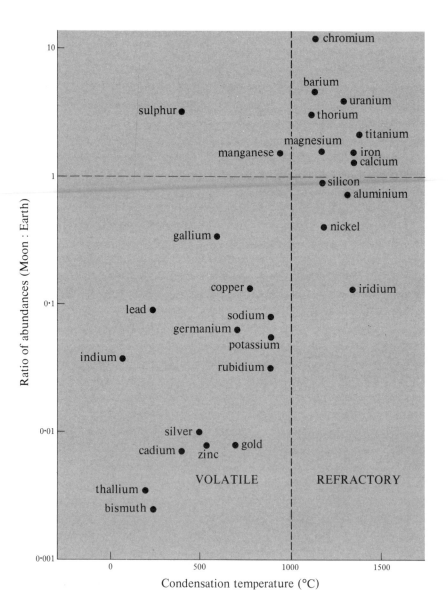

Figure 5.5 A comparison of element abundances in lunar and terrestrial basalts. The vertical dashed line separates volatile and refractory phases which control the abundances of these elements (see Table 5.1). (After Taylor 1975.)

The exceptional enrichment of volatile sulphur is because troilite (FeS) has crystallised in lunar basalts, whereas, in the relatively oxygen-rich terrestrial volcanic environments, SO_2 gas would be formed and expelled, leaving the remaining iron combined with oxygen in crystalline magnetite (Fe_3O_4).

Is the characteristic geochemistry of lunar basalts a fundamental feature of the lunar interior source region, or is it an effect of basalt lava extrusion into the high vacuum above the lunar surface? Two lines of argument point to the former. First, as elements can be lost by volatilisation only if they are at, or very close to, the surface (Taylor 1975), significant loss is unlikely. Secondly, remarkably uniform ratios of volatile to refractory elements were found in various samples from each Apollo site. It is concluded that the source region for lunar basalts is enriched in refractory elements at the expense of volatile and siderophilic elements compared with the Earth's mantle. Similar conclusions have been made from studies of the anorthositic highland rocks that are rich in Al_2O_3, CaO and other refractory elements. *This fundamental geochemical lunar signature, which distinguishes them from similar terrestrial rocks, is common to all sampled lunar rocks.*

Apart from these geochemical results, the Apollo missions also made some relevant geophysical investigations. They deployed several seismic receivers, which have since recorded signals from the deliberate impacts of expended space ironmongery, from natural impact events and from moonquakes that occur at the rate of about 3000 per year. The structural information deduced from these signals is illustrated in Figure 5.6. The strongest seismic discontinuity on the Moon occurs at 60 km depth where P-wave velocities change from $7 \, \text{km s}^{-1}$ (thought to characterise the anorthositic gabbro/basalt crust) to $7 \cdot 7 \, \text{km s}^{-1}$ for the lunar 'mantle' which is probably peridotite (see Ch. 7). P-wave velocities increase progressively from $7 \cdot 7$ to $8 \cdot 3 \, \text{km s}^{-1}$ down to 1000 km depth and S-waves are also recorded in this depth range. The region down to 1000 km is therefore recognised as the lunar 'lithosphere'. Below 1000 km, P-wave velocities drop by $0 \cdot 3 \, \text{km s}^{-1}$ and S-waves are attenuated, suggesting no startling change in composition or phase, but merely the introduction of some melt. The whole central region, radius 700 km, may be partially molten, and is termed the lunar 'asthenosphere'. The seismic evidence has not yet shown whether the Moon has a metallic core, though some workers claim palaeomagnetic evidence for its existence.

Heat flow measurements (see Note 9) on the Moon yielded values about half those on the Earth (see Ch. 8) – rather higher than expected as the Moon has a large surface-area/volume ratio and should have lost internal heat and cooled more rapidly than the Earth. To obtain the observed heat flow in a Moon very depleted of potassium, and hence radioactive ^{40}K (Fig. 5.5), it has been suggested that U and Th concentrations must exceed their relative solar abundance. This is more evidence (see Fig. 5.5) that refractory lithophilic elements are enriched in the Moon compared with the Earth and chondrites.

It is fair to ask what light this evidence casts on the various accretion models. The low concentrations of both volatile and siderophilic elements (Table 5.1) suggest that the Moon has concentrated high-temperature condensates – like a larger, but impure, version of the Allende inclusions. Even if the Moon does have a metallic core, its Fe/Si ratio must be lower than that of the Earth, and, together, these pieces of evidence could indicate heterogeneity in the Solar Nebula during planetary accretion. This conclusion would be particularly appropriate if the Moon formed independently and was captured by the Earth; however, the popular view is that the two bodies formed together, probably with the Moon accreting from a 'sediment' ring around the Earth (Ringwood 1975). In that case, as G. M. Brown (1978) has emphasised, the bulk composition of the Moon may resemble that of the Earth's mantle, after the loss of volatile and siderophilic elements. If the Moon did form from a ring of volatile-depleted silicate matter around the accreting Earth which was also developing a siderophilic element-enriched core, then the geochemistry of the two bodies together can be explained in terms of an initially homogeneous chondritic mass. The mass either failed to accrete completely or lost some material after accretion. The mass of the Moon is only 1·2% that of the Earth, and so its removal would hardly affect estimates of the Earth's initial composition (Section 5.4.1). Therefore, until we are more certain about the origin of the Moon, our knowledge of its geochemistry cannot be used to discriminate conclusively between accretion models. But together, the Earth and the Moon may have a bulk chondritic composition depleted in all volatile materials including alkali elements.

The remaining subsections describe the little evidence from Mars, Venus and Mercury that is relevant to planetary accretion models. These planets are potentially very relevant but the sparse available data lead to rather complex arguments. Therefore, these subsections have been boxed; the interested reader may follow the arguments which are summarised in Section 5.5.

5.4.3 *Mars*

The close-up and surface pictures of Mars returned by the Mariner and Viking missions of the 1970s showed that the planet has a spectacularly diverse topography with vast dry river channels, lava flows and volcanoes. Unfortunately, our knowledge of the ages of these features is based on highly debated crater density results. But it is agreed that Mars had a long history of activity, which probably ceased sometime during the last 2000 Ma. The Viking landers of 1976 made some crude x-ray fluorescence analyses of surface samples, which revealed heavily weathered silicates, oxides, sulphates and carbonates based on the overall chemistry of basalt, possibly more iron-rich than terrestrial basalts. Higher concentrations of the volatile element, potassium, in near-surface rocks were inferred from the much higher $^{40}Ar/^{36}Ar$ ratio of the Martian atmosphere compared with the Earth (radioactive decay of $^{40}K \rightarrow {}^{40}Ar$ has increased this ratio from a presumed common primitive Solar Nebula starting value). But there are other explanations (see J. V. Smith 1979).

Geophysical studies have greater potential for defining the internal structure and bulk composition of the planet, but the Viking seismic experiments were an almost total failure. This leaves us with a moment of inertia and a mean density ($3940 \, kg \, m^{-3}$) which is just over 70% of that of the Earth. But this need not imply a major difference in composition because, being so much less massive than the Earth, its material will be less compressed. (Pressure inside a planet is a function of depth and, since Mars has a radius about half that of the Earth, the same material at the centre of Mars would have a much lower density than at the centre of the Earth.) Mars must have a core because its polar moment of inertia ($C=0·376$.) is less than that for a solid sphere of uniform material ($C=0·4$). However, the Martian core is either *smaller* or less dense than the Earth's (where $C=0·331$ – see discussion by Cole (1978)). Mars and the Earth do not appear to have markedly different compositions (see also Section 5.5), and the less effective partitioning of iron into the Martian core correlates with its abundance in surface rocks. This might indicate a more oxidised composition for the planet, giving more iron in the lithophilic layer, a smaller core than for the Earth (see Section 5.3) and less accretional loss of

88

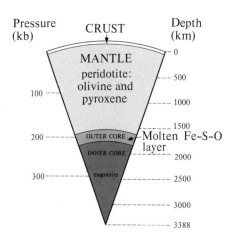

Figure 5.6 Schematic illustration of the lunar interior based on seismic evidence and including probable rock types in each layer.

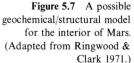

Figure 5.7 A possible geochemical/structural model for the interior of Mars. (Adapted from Ringwood & Clark 1971.)

volatiles from an otherwise chondritic planet. To say more, at this stage, requires the use of models of the internal structure and composition of Mars, and it is hardly fair to use geochemical models to test accretion models! One such geochemical model, based on an extremely oxidised bulk composition – with an Fe_3O_4 core – is shown for interest in Figure 5.7, but models based on an Fe–FeS core may be equally valid (for example, Johnston and Toksöz 1977).

A more volatile-rich (oxygen and possibly potassium, see above) composition for Mars is not sufficiently discriminating to be decisive about accretion models. On the heterogeneous model, we should expect a greater concentration of volatiles in Mars than in the Earth, because it is the most distant terrestrial planet from the Sun. And, on the homogeneous model, it can be argued that this relatively small planet will have lost less volatiles during accretional heating.

5.4.4 *Venus*

The two remaining terrestrial planets are less well explored by space missions and, unlike Mars, neither Venus nor Mercury has a moon, so that even their moments of inertia cannot be determined (see Section 3.3.2).

From radar measurements, which penetrate the very thick atmosphere, we know that the Venusian surface has canyons, rift valleys, craters and possibly volcanoes, but not enough is known to test the geological speculation that such a large planet (82% of the Earth's mass) should still be active (see Kaula 1975). Gamma-ray measurements of K, U and Th during the Venera fly-by missions provide the only chemical data from the surface of Venus. The measured abundances of these elements range between the values for terrestrial granites and basalts, and with similar K/U/Th ratios to those on the Earth, suggesting similar volatile (K) to refractory (U and Th) element ratios for Venus and the Earth.

Venus has a mean density of $5240 \, kg \, m^{-3}$, 5% less than for the Earth but, as Venus is slightly smaller, the deep interior should be less compressed and less dense. Assuming the internal structures and compositions of the two planets are otherwise identical, Ringwood and Anderson (1977) calculated that the actual density difference of 5% should be reduced to 1·7% after allowing for compression. Thus, Venus is unlikely to have a higher Fe/Si ratio than the Earth, predicted by the heterogeneous accretion model, since Venus should then have a higher density than the Earth, after correcting for pressure. Lewis (1972), an advocate of heterogeneous planetary accretion, argued that, if Venus accreted at a slightly higher temperature than the Earth, such that the only difference between the two planets is the relative absence of troilite (FeS) in Venus, then Venus would have a lower density than the Earth. (Troilite is both heavier and relatively more volatile than silicates – Table 5.1 – hence, Lewis was suggesting that troilite condensed in the accretion of the Earth, but not in that of Venus.) However, there are two difficulties with this argument: (i) the Venusian atmosphere contains sulphur particles and sulphuric acid, presumably outgassed from volcanoes, so the planet does contain sulphides, and (ii) Ringwood and Anderson (1977) showed that Lewis's model also necessitates that Venus has less oxygen relative to the Earth, a feature that would increase rather than reduce the density contrast. (Fe_3O_4 is more volatile than FeS; Table 5.1.) Instead, Ringwood and Anderson proposed that Venus has lost slightly less volatiles, including oxygen, than the Earth, precisely as predicted by the homogeneous accretion model (Table 5.2), but that otherwise the two planets are similar. To remove the 1·7% density difference, their model for the Venusian interior has a smaller core, 23% of the planetary mass, than the Earth and more oxidised iron in the mantle.

5.4.5 *Mercury*

Space probes that have visited Mercury record a heavily cratered and apparently inactive planet. This is not surprising in view of the small size of this planet as heat losses are greatest from small planetary bodies (see Section 5.4.2).

Mercury is about 15% too dense ($5420\,kg\,m^{-3}$) to be made of similar materials to the other planets. Volatilisation of all chondritic materials up to alkali feldspars (Table 5.1) is even insufficient to account for this density and smaller amounts of mantle-forming silicates (olivines and pyroxenes) and more iron-rich core materials must occur in Mercury than in the other planets. So the density of Mercury provides the best evidence for variable Fe/Si ratios in the terrestrial planets, and most workers have followed Urey (1952) who proposed a high content of metallic iron to account for Mercury's density. To obtain a satisfactory model for the interior of Mercury a metallic core, approximately 1800 km in radius, surrounded by a further 600 km of refractory-element-enriched silicate mantle is postulated.

Mercury is the closest planet to the Sun, where temperatures in the pre-accretion nebula were highest among the terrestrial planets: this evidence strongly supports heterogeneous accretion (Table 5.2). Mercury could not have attained its present composition by volatile loss during accretional heating of chondritic material. However, it is possible that Mercury accreted in its own 'zone' of a heterogeneous nebula where most of the iron had condensed and just after magnesium silicates started to form (for a review, see J. V. Smith 1979).

5.5   Synthesis: a compromise accretion model

From the survey of possible planetary compositions in Section 5.4, we can extract the following points:

(a)   The increase in mean density with size (Table 4.1) for the planets Mars, Venus and Earth can be explained entirely in terms of two factors. First, the increasing state of internal compression for larger planets, which results in an increase in mean density with no change in composition. Second, however, it is necessary to infer slight increases in the proportion of light, volatile elements, particularly oxygen, in Venus and especially in Mars to account for surface rock analyses and mean density data. If Mars and Venus have more oxygen than the Earth, then their metallic cores will be correspondingly smaller fractions of their masses.

(b)   Because the Earth–Moon system has a composition that matches well the composition of carbonaceous chondrites, except for the loss of the most volatile elements, it follows that Mars, Venus and the Earth are approximately chondritic

90

at least in the abundances of refractory elements. This suggests that they all accreted from similar, probably homogeneous, material and that volatile losses were influenced by accretional heating since the smaller the planet, the smaller the amount of volatile material that has been lost. (Section 5.2.2).

(c) Figure 5.8, a plot of planetary densities and radii, emphasises the correlation for the Earth, Venus and Mars, and points to the very different composition that must apply to Mercury. The only relatively dense material that can be present, and which had sufficient abundance in the primitive Solar Nebula, is iron. So, Mercury comprises higher-temperature condensates than the other terrestrial planets, and heterogeneous mixtures of condensed grains must have existed in the Solar Nebula prior to planetary accretion. This is emphasised by the compositions of the outer planets and certain chondrites which contain high-temperature condensates.

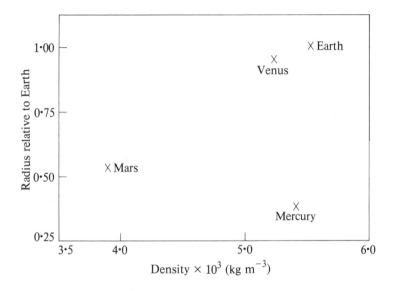

Figure 5.8 Plot of radius against density for the terrestrial planets, showing that density is an important function of size, hence internal compression, for Mars, Venus and the Earth.

Here is one possible model for planetary accretion starting from the situation shown in Figure 5.9. The accumulated evidence from all the planets and meteorites shows that the Solar Nebula was heterogeneous before planetary accretion. However, the evidence from Mars, the Earth and Venus indicates that, within the nebula, there was a homogeneous 'zone' of grains that formed these three planets. A minimum of three zones for the Solar Nebula is indicated by the evidence from all the planets, based on progressively lower-temperature condensates further from the centre. Apart from the 'chondritic zone', where temperatures must have been about 100°C when the Earth formed (Table 5.1), there was an inner, high-temperature zone in which Mercury accreted and an outer, low-temperature zone ($<0°C$) from which the vast bulk of the outer planets developed. According to this model, the Earth accreted in the middle zone by the physical processes described in Section 4.3. Our planet heated up, lost volatiles and formed a metallic core as described in Sections 5.2.2 and 5.2.3.

Attention is now focused on the Earth, and in the remaining chapters we look in more detail at the physics and chemistry of the core, the mantle and the crust.

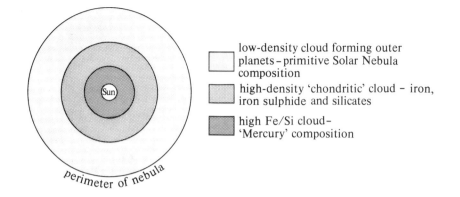

Figure 5.9 Schematic illustration of possible zones comprising condensed grains of different compositions in the Solar Nebula immediately prior to planetary accretion. See text for further details.

Summary

1 The chemical elements inside the terrestrial planets and in meteorites ·are segregated into three overlapping classes – lithophilic (crust and mantle silicates), chalcophilic (sulphide phase) and siderophilic (metallic phase) – according to their Periodic Table characteristics expressed through electronegativities (Table 5.3).

2 A comparison of the Earth's crust with the heavy-element composition of the primitive Solar Nebula (solar spectra) indicates enrichments in lithophilic elements, but depletions in siderophilic and chalcophilic elements. However, the bulk composition of the Earth is estimated to be closely similar to the heavy elements in solar spectra and, in comparison with carbonaceous chondrites, the Earth has lower concentrations of volatile components. Loss of H_2O, CO_2, etc., with the consequent loss of oxygen, means that much terrestrial iron is chemically reduced and forms a separate core rather than being combined with silicates.

3 There are two principal ways in which fundamental chemical sorting may have been accomplished in the terrestrial planets (see Table 5.2):
   (a) By element segregation in the Solar Nebula according to condensation temperatures (Table 5.1), leading to *heterogeneous* planetary growth of refractory cores, then less refractory mantles as temperatures fell in the region of each accreting planet.
   (b) Internal element segregation by melting processes during or after accretion, possibly from *homogeneous* ready-condensed nebula material.

4 The evidence from surface rock chemistry, seismic data, density and moment of inertia values support the Earth, Venus and Mars having similar chondritic bulk compositions which vary only by increasing contents of volatile elements, particularly oxygen, with decreasing size of planet. This suggests, as in the homogeneous accretion model, that volatilisation is a function of size and is caused by rising temperatures during accretion, due to the conversion of impact energies, and/or by the decay of short-lived radioisotopes. These effects are most significant for large planets such as the Earth. In that case, core formation in the Earth was a short-term runaway process that occurred early in Earth history.

5   Since Mercury probably has a more refractory, iron-rich composition and the outer planets have more volatile-rich (H, C, N, O) compositions than the 'chondritic' planets, it is suggested that at least three chemically distinct zones of condensed grains existed in the Solar Nebula before the planets accreted (see Fig. 5.9). The Earth formed in the middle zone from condensates at about 100°C, subsequently heated and became chemically layered as described above.

Further reading

General journal:
    Kaula (1975): 'Seven ages of a planet' – comparative planetology; history of internal processes and temperatures.
General books:
    Cole (1978): physics of the planets.
    Taylor (1975): post-Apollo review of the Moon's origin and geochemistry.
    Ernst (1969); Mason (1966); Turekian (1972): aspects of geochemical theory and the chemical history of the Earth.
Advanced journals:
    Grossman & Larimer (1974): early chemical history and condensation temperatures of the Solar System.
    Clark et al. (1972): heterogeneous accretion of the Earth.
    Ringwood & Anderson: (1977): comparative studies of the Earth and Venus.
    J. V. Smith (1979): broad review of planetary mineralogy and geochemistry.
Advanced book:
    Ringwood (1975): homogeneous accretion theory.

6 The Earth's core

6.1 Problems posed by the core

So far we have introduced the principles that are to guide our understanding of the Earth's internal constitution. This is the first chapter that combines this knowledge to examine a specific region of the Earth. We know already that the core has a much higher density than the Earth as a whole and that it consists of two parts: a liquid outer core surrounding a solid inner core. What, then, are the problems to be solved?

First, it is useful to determine the composition of the inner and outer cores, and then to consider why the inner core is solid despite its presumed higher temperature compared with the outer core. The relevant information to solve these problems comes from several sources. In Chapter 3, the density of the core was deduced and this has to be matched with a suitable composition at high pressure. Chapters 4 and 5 introduced the Chondritic Earth Model and suggested that, generally, the Earth may have a bulk composition close to that of chondritic meteorites. Section 5.3 discussed the ways the chemical elements were separated after accretion according to their electrochemical properties. The evidence indicates quite clearly that the core must consist predominantly of iron, with some other elements to provide a closer match with the density and other observed properties.

A new source of information, specific to the core, is the existence of the Earth's magnetic field which, it will be shown, is generated in the outer core by a kind of 'dynamo' action, driven by motions in the liquid. In turn, this requires an energy source.

Thus, any model of the core must have compositions for the inner and outer core that (i) are consistent with their known densities and the laws of geochemistry; (ii) has a temperature distribution that permits the inner core to be solid within a liquid outer core; and (iii) has a sufficient source of energy to generate the magnetic field.

The most simple model, postulated by Jacobs (1953), comprises a chemically homogeneous core whose melting temperature increases with depth, due to the increasing pressure, faster than the actual temperature. This situation is shown in Figure 6.1a, where the melting temperature drops across the core/mantle boundary because of the change from a predominantly silicate to an iron-rich constitution. Provided that the melting-point curve for the core material is steeper than the temperature gradient, then there will be a change from liquid to solid which will define the inner core/outer core boundary. This explanation was unchallenged until 1971, when it was suggested that the thermal gradient in the core must be at least adiabatic to maintain the convection needed to produce the magnetic field (Section 6.2). Higgins and Kennedy (1971) calculated that the adiabat would be *steeper* than the melting-point curve (Fig. 6.1b). In other words, earlier estimates of temperature variation in the core (curve B) were subadiabatic and curve A represents the minimum thermal gradient for thermal convection to occur in the core. However, the intersection of the melting-point curve for a homogeneous core with curve A

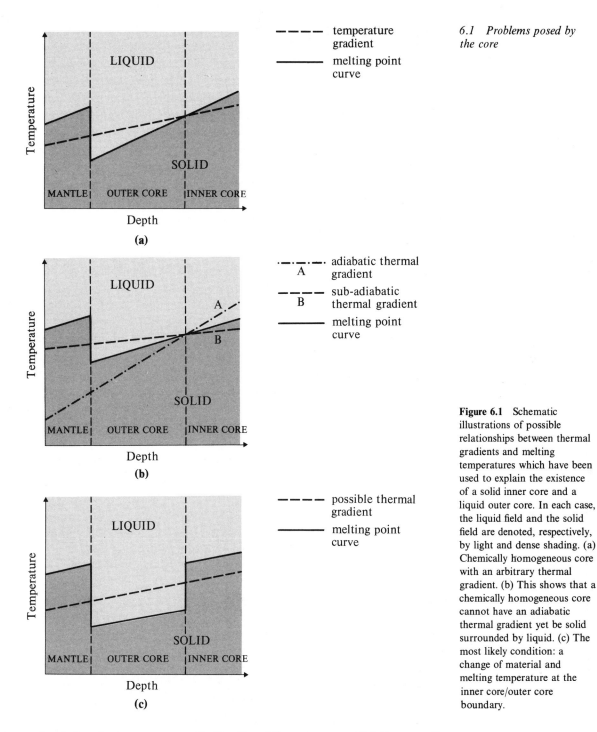

Figure 6.1 Schematic illustrations of possible relationships between thermal gradients and melting temperatures which have been used to explain the existence of a solid inner core and a liquid outer core. In each case, the liquid field and the solid field are denoted, respectively, by light and dense shading. (a) Chemically homogeneous core with an arbitrary thermal gradient. (b) This shows that a chemically homogeneous core cannot have an adiabatic thermal gradient yet be solid surrounded by liquid. (c) The most likely condition: a change of material and melting temperature at the inner core/outer core boundary.

shows clearly that the inner core must be liquid and the outer core solid, the opposite of what is observed.

Here the earlier assumption of a homogeneous core broke down and the more likely situation is shown in Figure 6.1c. It was realised that the inner and outer cores probably have different chemical compositions and, thereby, different melting temperatures. If the inner core is solid because it has much the higher melting

95

temperature, then the required adiabatic temperature gradient can be maintained across the fluid outer core. Later, it will be argued from density data that the inner core contains a significant amount of nickel, whereas iron in the outer core is diluted with sulphur which greatly reduces its melting point. Also, it will be shown that the energy required to account for convection and, hence, the magnetic field, comes either from the radioactive decay of ^{40}K, dissolved along with the sulphur in the outer core, or, alternatively, from the separation of iron–nickel alloy into a growing inner core, which is equivalent to the settling of a denser phase, with consequent release of gravitational energy. In both cases, the temperature gradient across the outer core is likely to be close to adiabatic, and this can be used to set limits on the likely temperatures within the core and on the heat flowing from the core to the mantle.

6.2 The Earth's magnetic field and the problem of energy

6.2.1 *A description of the field*

Most people regard the Earth's magnetic field merely as a useful means of navigation. Those who recollect experiments with iron filings might suggest that the field is much like that of a bar magnet. To a good approximation, the field at the Earth's surface *is* like that due to a powerful bar magnet (i.e. a **dipole**) situated at the Earth's centre and aligned roughly along the axis of its rotation (Fig. 6.2). Since the Earth's core is likely to be made largely of iron, it might be thought that the solid inner core behaves like a permanent magnet. However, there are several reasons why matters are not so simple, and these are now given.

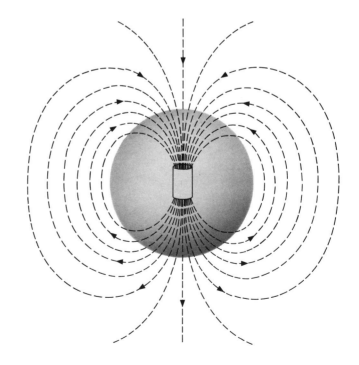

Figure 6.2 To a first approximation, the Earth's magnetic field is like that due to a strong bar magnet at the centre of the Earth. The arrows indicate the direction of lines of force (dashed), and the angle at which these lines cut the surface is the **inclination**. Thus, inclination is 0° at the magnetic equator and 90° at the magnetic poles of a dipole.

96

(a) The temperature of the core undoubtedly exceeds the low-pressure **Curie temperature** of all known magnetic alloys or minerals (usually 500–800°C), above which solids lose their permanent magnetisation.

(b) Although the Earth's field is approximately that of a dipole, there are significant departures. As Gauss first showed in about 1835, there is a small **non-dipole** component comprising apparently random highs and lows. The non-dipole field could be interpreted in terms of a permanent magnet with non-uniform magnetisation. However, this is disproved by the next piece of evidence.

(c) The Earth's magnetic field, both the non-dipole and the dipole components, changes position and strength with time. The sum of all these changes on time scales from tens to thousands, and even millions, of years is known as **secular variation**. For instance the direction of the magnetic field with respect to geographic co-ordinates changes slowly so that navigators have to make a correction to their compass bearings. The correction, measured as **declination** (Fig. 6.3), applies to conventional horizontally mounted compass needles which respond only to the horizontal component of the Earth's field. However, compass needles mounted in the vertical plane also change their angle of **inclination** (i.e. the angle at which the lines of force cut the surface – Fig. 6.2) with time, as shown in Figure 6.3. The field changes with time all over the Earth's surface are the product of variations in the direction and strength of both the dipole field and the non-dipole field. There are two superimposed departures from a simple axial dipolar magnetic field. (i) The dipole field is not aligned along the rotation axis, but is inclined at a small (usually < 12°), but variable, angle. As the angle varies, the magnetic poles seem to 'wobble' around the geographic poles, though the field is axial and dipolar on average. Dipole wobble is the major cause of the variations over the past few centuries shown in Figure 6.3. (ii) The non-dipole field locally adds to, or subtracts from, the dipole field.

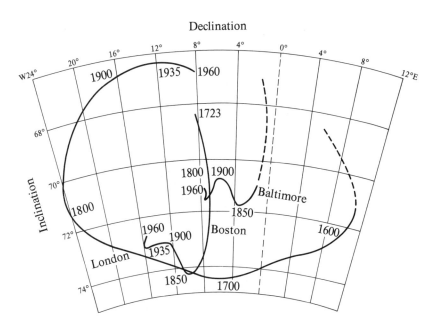

Declination

Figure 6.3 Variation of the direction of the Earth's magnetic field as recorded by various observatories in recent centuries. Declination is the horizontal angle between magnetic and geographic north, while inclination is the angle below the horizontal at which a vertically mounted compass (dip-circle) comes to rest (see Fig. 6.2). (Source: Jacobs 1975. Reprinted with permission from Jacobs, J.A. 1975. *The Earth's core.* © Academic Press Inc. (London) Ltd.)

To extend magnetic field records back into geological time, studies of **palaeomagnetism** (or 'fossil magnetism') in ancient rocks are used. This technique is possible because many rocks acquire some permanent magnetism from the field existing at the time they formed. For example, the iron-rich minerals in a basalt lava cooling through their Curie temperatures will become magnetised. This means that the rocks may retain evidence of the former magnetising field until the present day. This property can be exploited by drilling oriented core samples from a rock outcrop, followed by magnetic measurements in the laboratory to determine palaeopoles. (The use of palaeomagnetism to determine past continental positions will be discussed in Chapter 10.) By applying this technique to lava sequences which have been erupted over millions of years, it has been found that the whole field has *reversed* many times in the past.. For example, north palaeopoles become south palaeopoles and vice versa. Palaeomagnetic records of reversals for the past 80 Ma are shown in Figure 6.4. Further back in time, it seems that reversals have occurred throughout much of geological history (for further details see McElhinny 1973). Moreover, the strength of the field also varies considerably, especially during reversals which are characterised by minima of field strength (Dunn *et al.* 1971).

Clearly, the Earth's magnetic field is a highly variable and dynamic system on all scales of time, a system that has been in existence for most of the Earth's history and which is not due to permanent magnetism in the core. How, then, is the magnetic field generated in the Earth's core?

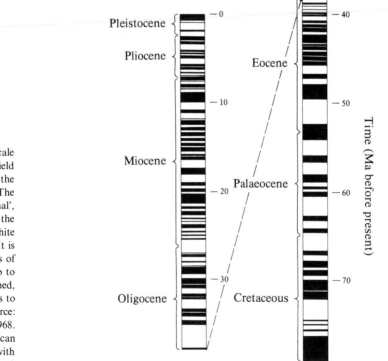

Figure 6.4 Polarity time scale for the Earth's magnetic field (dipole component) over the past 80 million years. The black intervals are 'normal', i.e. in the same sense as the present field, and the white intervals are reversed. It is thought that the process of field reversal may take up to 10^4 a but, once established, each polarity period seems to last about 10^5–10^6 a. (Source: Heirtzler *et al.* 1968. Copyright American Geophysical Union, with permission.)

98

6.2.2   The geomagnetic dynamo

The changes in the position and strength of the magnetic field suggest a source in the fluid, mobile outer core. Almost all of the solid mantle and inner core are above their Curie temperatures, and so these regions are unlikely to have any permanent magnetism. Nor are motions within them sufficiently vigorous to generate the observed secular variation. This leaves electrical currents in the outer core as the only plausible means of generating the magnetic field. The idea is that there are current loops, very roughly as in a solenoid or electric coil, by which the various components of the magnetic field are generated, and this has led to the widely accepted 'geomagnetic dynamo' model (see Busse 1975 for a technical summary).

Figure 6.5 shows a simple disc dynamo analogy. In Figure 6.5a, the disc, which is a conductor, rotates in a magnetic field, an e.m.f. is generated between the axle and rim but no current flows. In Figure 6.5b, there is an external circuit carrying a current from the axle to the perimeter of the disc. This current produces a secondary magnetic field and it requires an external magnetic field and an energy source to provide the rotation. In Figure 6.5c, the external magnetic field, or exciting field, of the dynamo has been dispensed with and, instead, the current is passed through a coil around the axle of the disc so that the secondary magnetic field it produces is used for its own excitation. This is now a **self-exciting dynamo** which will continue to generate a field so long as it is rotated. It provides a most useful analogy for the generation of the Earth's magnetic field by current loops in the core: once the dynamo system gets underway with a small initial magnetic field (perhaps that of the Sun) it requires only an energy source to be maintained. If the current, but not the direction of rotation, is reversed, the field also reverses. The process of **self-reversal** of the Earth's field is undoubtedly more complex than this, but it has been successfully modelled using a pair of dynamos that generate fields for each other (see Jacobs 1975, p. 153, for a discussion). In the liquid outer core, there may be an almost infinite number of mutually interacting current loops, with different shapes and forms, which grow, fade and reverse on all scales of time. Perhaps the current loops are constrained by the Earth's rotation, so accounting for the approximate north–south alignment of the magnetic field.

Estimates have been made of the total continuous energy input required by the geomagnetic field. These range from 10^9 to 10^{11} W (watts), depending on the efficiency with which thermal or mechanical energy is converted into magnetic energy (Lowes 1970 and Note 8). In comparison with the power released from earthquakes (about 10^{12} W) and with heat flowing from the surface of the Earth

Figure 6.5 Illustration of the dynamo model for the origin of the Earth's magnetic field. In (a) the disc is rotating in a magnetic field, a potential difference is generated, but no current flows. In (b) the external circuit enables current to flow, and in (c) the current supplies a field and causes the dynamo to be self-exciting. For further details, see text. (After Bullard 1971, from *Understanding the Earth*, Gass, I.G., Smith, P.J. & Wilson, R.C.L. (eds), 71–80. Sussex: Artemis Press.)

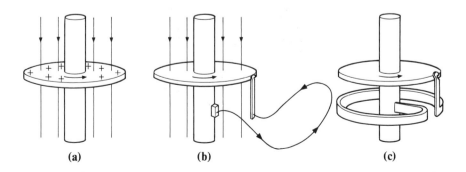

(a)   (b)   (c)

$(3–4 \times 10^{13}$ W), that used by the magnetic field is relatively small. However, there are problems in making available, in mechanical form, the requisite energy at such great depths.

6.2.3 *Heat and energy in the core*

First, **thermal convection**, due either to radioactive decay or to the release of latent heat during solidification and growth of the inner core, will be considered. This form of energy must be capable of maintaining a thermal gradient which is at least adiabatic in order that thermal convection may occur.

The only long-lived radioactive isotope capable of providing the required energy which may occur in the core is ^{40}K. All the alkali metals (K, Rb, Cs, etc.) form large atoms at low pressures, and hence are likely to be lithophilic and concentrated in the crust (Section 5.3), but both thermodynamic theory (Lewis 1971, Hall & Rama Murthy 1971) and experimental data (Bukowinski 1976) indicate that the single outermost s-electron may transfer to the smaller d-shell at high pressures, giving these elements chalcophilic properties. This means that they may have become incorporated into the major sulphide-bearing zone of the Earth, i.e. the outer core. (Incidentally, the idea receives some support from the occurrence of a potassium sulphide mineral, djerfisherite, in meteorites.)

On the assumption that potassium is lithophilic in the Earth, in Section 5.4.1 we concluded that the Earth has only about one-third of these alkali metals compared with those in chondrites. If all the 'deficient' ^{40}K (half-life $= 1300$ Ma) were present in the outer core, then about 10^{13} W would be produced – *apparently* quite adequate to 'drive' the geomagnetic dynamo. However, Ringwood (1977b) and others contend that the Earth has lost potassium relative to chondrites and that, at most, the core contains only *traces* of potassium. To generate 10^{11} W, the outer core would need to have 5 parts per million of potassium, which is 1.5% of the potassium budget of a chondritic Earth. However, this energy would need to be converted with near 100% efficiency into mechanical energy to provide an adequate source for the magnetic field. As will be explained below, this may be far from the truth!

An alternative source of heat energy in the core (Verhoogen 1961, 1973) is the latent heat released by progressive growth of the solid inner core. It is argued that the outer core is crystallising progressively by addition of iron–nickel alloy as the temperature falls, giving rise to growth of the inner core. In the process, some of the heat already present in the Earth, which caused its core to become molten, was made available within the core throughout the Earth's history. Verhoogen estimated that, with a cooling rate of $12°$C every 10^9 years, heat is released at a continuous rate of 10^{12} W – again, apparently of the right order to generate the magnetic field.

Although sufficient power to drive the geomagnetic dynamo may be derived either from radioactivity or latent heat there is the question of how useful this heat energy is in transporting the core materials mechanically; that is, the *efficiency* of conversion to mechanical motion. To generate thermal convective motions in the outer core, the rate of heat production must be so great that heat cannot be transported to the core/mantle boundary simply by conduction down the adiabatic gradient. Metchnik *et al.* (1974) calculated that, to achieve convection, the minimum heat production should be $2·5 \times 10^{12}$ W. Below this value, the conversion of heat energy into magnetic energy by a dynamo has almost zero efficiency because there is no material transport. The efficiency rises to a useful level of a few per cent as thermal power approaches 10^{13} W, in which case mechanical power for the dynamo in excess of 10^{11} W becomes available (Stacey 1977). Therefore, only if a substantial part of the Earth's potassium is in the core (about 0.1% as K, equivalent to 67% of the potassium in a chondritic Earth), will sufficient heat energy to drive the magnetic dynamo be available. Because of the low efficiency of a thermally driven dynamo, the power output due to smaller concentrations of ^{40}K or due to inner core crystallisation is probably inadequate.

6.2.4   *Gravitational energy in the core*

Gravitational forces may operate in two different ways to induce motion in the core. In the first, the torque produced by the Moon on the Earth's equatorial bulge (see Fig. 3.6) stirs the liquid outer core. However, the energy available from these precessional torques would be only about 10^7 W, which is inadequate to generate the magnetic field (see Jacobs (1975, p. 175) and Loper (1975) for discussion and further references).

The second gravitational mechanism depends on forces entirely within the core and, currently, is the favoured mechanism. As with the earlier theory of latent heat release (Section 6.2.3), this mechanism relies on the progressive growth of the solid inner core by cooling and crystallisation of a relatively dense alloy in the outer core throughout the history of the magnetic field. It is argued that material solidifying onto the surface of the inner core is richer in nickel than the average composition of the outer core. To achieve this, a layer depleted in nickel must form outside the inner core. As it is *less dense* than the rest of the outer core, it will be gravitationally unstable and so will tend to rise and produce convective circulation. The solid iron–nickel material that crystallises in the outer core will be *more dense* than the remaining material and so will sink towards the inner core. The sum of these processes is known as **gravitational convection** (postulated by Gubbins (1977) and Loper (1978)).

Detailed estimates of the likely density contrast between the solid and buoyant liquid (Loper 1978) lie in the range 500–2500 kg m^{-3}. Loper estimated that the growth of the Earth's solid inner core from a totally liquid core would yield 5×10^{28} J (joules) in this way. This represents a steady 3.6×10^{11} W, assuming a uniform rate of release over the Earth's history. But most theorists consider that the inner core is crystallising *less rapidly* than early in its history. Even so, the power available from gravitational convection is likely to meet the needs ($10^9 - 10^{11}$ W) of the magnetic field. *The efficiency of the process must be high because it results directly in material transport.*

The formation and growth of the inner core in this way also releases *latent heat* energy (Section 6.2.3) and, additionally, the passage of electric currents through the core material generates heat by resistive dissipation (see Note 8). Both forms of heating may assist gravitational convection by adding a thermal component and so maintaining the high efficiency of the process (near 100%).

To summarise, at present there are two plausible mechanisms whereby the Earth's magnetic field may be generated. First, the decay of radioactive ^{40}K may produce *thermal* convection, but, with its low efficiency, relatively large amounts of ^{40}K are needed. This would tend to prevent cooling of the core and hence prevent significant growth of the inner core. On the other hand, growth of the inner core, due to slow cooling, produces much more efficient *gravitational* convection which does not require that the temperature gradient is adiabatic. However, given the supply of heat from inner core crystallisation, resistive heating, etc., the thermal gradient may well be close to adiabatic even if convection is due to gravitational processes.

6.3 The constitution and evolution of the core

6.3.1 *Density and composition*

The density of the inner core is known only approximately (Ch. 3) but a range of $12\,600$–$13\,000\ kg\ m^{-3}$ is currently accepted and, according to shock-wave experiments at the appropriate pressure (about 3·6 Mbar, see Fig. 3.14), this is too dense for pure iron. The only available diluent of suitable abundance and density is nickel, which implies an Fe–Ni alloy as in iron meteorites. The amount of nickel in the inner core is not known because of the imprecision of density determinations but, on the meteorite analogy, it is likely to be 10–20%.

Whereas the inner core comprises only 1·7% of the Earth's mass, the outer core is much larger and more massive (30%) and its density is much better known. Since it must be well mixed by the motions that produce the dynamo, it must also be fairly homogeneous. Density values range from about $9900\ kg\ m^{-3}$ at the core/mantle boundary to about $12\,200\ kg\ m^{-3}$ at the higher pressures of the inner core/outer core boundary (Fig. 3.14). Shock-wave data for 1·4 Mbar (the core/mantle boundary) show that pure iron has a density of roughly $10\,600\ kg\ m^{-3}$. Hence, pure iron is too dense for the outer core – the converse of the inner core – and a lighter diluent must be present. (It is this evidence that points unequivocally to the chemical zoning of the core discussed in relation to Figure 6.1c.)

There are only two likely diluents for iron in the outer core which are sufficiently abundant in the Earth: silicon and sulphur. Oxygen and magnesium (Fig. 5.3) are unlikely contenders for the outer core because of their strong lithophilic tendencies (Table 5.3). Sulphur is now the favoured outer core diluent, but it is useful to recount some of the arguments used during the 1960s in favour of silicon. A main point is that silicon is lighter than sulphur (atomic weights: Si, 28; S, 32) and so less silicon is needed to provide a density match. Shock compression studies at outer core pressures (Birch 1968) showed that a 90% Fe, 10% Si alloy yields a good match with the density observed, whereas nearly 15% sulphur is needed. Bearing in mind that a chondritic Earth would contain about 8% sulphur and that the outer core comprises roughly one-third of the Earth's mass, more than half of the sulphur would have to be concentrated in the core. Formerly, this was taken as evidence against sulphur, but it is now thought that sulphur may have been efficiently separated into the core by chemical processes in the early Earth (Section 5.3). Moreover, estimates of the sulphur content of the outer core have been revised downwards (see below). Support for the Fe–Si outer core model was also derived from studies of enstatite chondrite meteorites which have minute amounts of silicon dissolved in their metallic phases.

However, since 1970, the case for an iron–sulphur outer core has gained popularity. The electronic configurations of the elements (Section 5.3) indicate that iron has a strong affinity to dissolve any sulphur during the process of core formation, leaving only the excess iron to form oxides or to be metallic. On the other hand, silicon, of similar electronegativity to iron, forms small ions which complex with oxygen in silicates, and which are then strongly lithophilic. Since iron oxide does occur in the crust and mantle, the presence of more lithophilic silicate complexes in the core would mean that there is considerable chemical disequilibrium across the core/mantle boundary. As this is unlikely, the case against silicon and for sulphur in the outer core is reaffirmed. Moreover, if the outer core does not contain much of the sulphur budget of a chondritic Earth, then the Earth must have lost large amounts of sulphur during accretion which (as argued for Venus in Section 5.4.4) would cause the Earth's density to be significantly less than observed (see Fig. 5.8). Finally, iron meteorites often contain considerable proportions of iron sulphide, as the mineral **troilite**, so that the balance of meteorite evidence also favours sulphur. The shock-wave experimental data of Ahrens (1979) show that a good match with the density of the outer core can be obtained with mixtures in the range *9–12% sulphur in iron*, with a preference for the upper end of this range. This case for sulphur in the Earth's outer core is persuasive, but perhaps the most convincing evidence comes from melting temperatures which are considered next.

6.3.2   Temperature and evolution

Figure 6.6a illustrates melting relations in the system Fe–FeS at atmospheric pressure. Whereas pure iron melts at 1539°C and pure FeS at 1230°C, *any* intermediate composition between these two extremes will become *partially* molten at a much lower temperature, 998°C. This is because the Fe–FeS system contains a **eutectic** at 75% FeS (corresponding to 27% S) where the first melts of any impure mixture will form. A eutectic composition mixture also becomes completely molten at 998%C. But, for any other mixture the amount of liquid increases as temperature rises until a complete melt is formed according to the heavy curves in Figure 6.6a. For example, a 20% FeS, 80% Fe mixture will start melting at 998°C and become

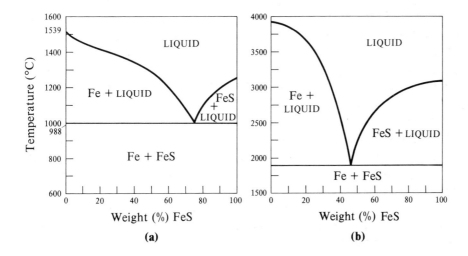

Figure 6.6 Melting relations of Fe–FeS mixtures (a) at 1 atm pressure and (b) extrapolated to the pressure of the core–mantle boundary (1·4 Mbar). (Data from Usselman 1975.) Note that the two temperature scales are different.

103

completely molten at about 1420°C. Clearly, the presence of sulphur considerably depresses the melting point of pure iron. Therefore, iron–sulphur mixtures in the outer core provide the low-temperature-melting mixture required to give the different melting points of the inner and outer cores illustrated in Figure 6.1c. In contrast, iron–silicon and iron–nickel mixtures do not form similar low-temperature-melting eutectics. So the inner core comprises a high-temperature-crystallising alloy, which explains its solid state.

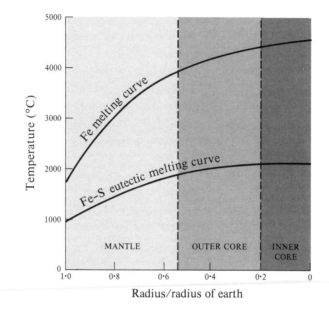

Figure 6.7 Variation of Fe and Fe–S eutectic melting curves in the Earth. Note from Fig. 6.6 that though the eutectic composition is 27% S (75% FeS) at the surface, it changes to 17.5% S (48% FeS) at the core/mantle boundary and to 15% S (41% FeS) at the inner core/outer core boundary. The two curves provide limits on the temperature in the outer core. (Data from Usselman 1975.)

To discover the melting relations of Fe–FeS systems at core pressures, Usselman (1975) conducted experiments on Fe–Ni–S mixtures up to 100 kbar and extrapolated the trends he found to higher pressures. Some of his results are shown in Figures 6.6 and 6.7; the temperature interval between the melting of pure iron and the eutectic increases with depth in the Earth. Therefore, during core formation (Fig. 5.2), *a eutectic melt* could sink more effectively through the mantle than pure iron, requiring a lower temperature gradient across the mantle to stay molten. As temperatures are likely to have increased with depth, perhaps due to gravitational energy release by the descending melt, a eutectic melt could scavenge the solid mantle and drain it of iron sulphide. On the other hand, the melting point of *pure iron* is only a little less than that of most silicates and, unless the early temperatures caused almost the entire Earth to be molten, there would be a tendency for a descending iron melt to freeze. So the descent of a near-eutectic FeS melt through semi-pervious solid mantle silicates provides a mechanism for core formation following homogeneous accretion (Section 5.2 and Fig. 5.2).

But does the outer core really comprise a eutectic mixture? Usselman (1975) showed that the amount of sulphur in the eutectic composition decreases with increasing pressure. He estimated that the amount of sulphur in the eutectic composition at core/mantle boundary pressures is 17·5%, decreasing to 15% (48 decreasing to 41% as FeS) at the inner core/outer core boundary and deduced that a few per cent of nickel in the outer core would have little effect on these temperature estimates. So the eutectic contains only a little more sulphur than the 9–12% inferred to be present from density data (12% S is equivalent to 33% FeS). The melting relations of Fe–FeS mixtures at core/mantle boundary pressures are shown schematically in Figure 6.6b, where the increase in temperature and decrease in sulphur content of the eutectic are shown. Limits on the core temperatures can now be deduced using Figures 6.6b, 6.7 and the extremes of composition. Accepting that the outer core is at the eutectic gives the *minimum* possible temperature at the core/mantle boundary to be 1800°C, increasing to 2100°C at the inner core/outer core boundary. The melting curve for pure iron gives *maximum* possible temperatures of 3900 and 4400°C at the top and bottom of the outer core. The inner core would melt if temperatures were any higher than these. (Jacobs (1975, p. 99) lists 16 independently derived values for iron melting at core/mantle boundary pressures. Their total range is 2340–4800°C with a *mean* at 3750°C, quite close to Usselman's (1975) value.)

According to the estimates given above, the temperature at the core–mantle boundary is between 1800 and 3900°C. Elsasser *et al.* (1979) discussed the probable temperature distribution in the core in relation to the way heat flow from the core affects deep mantle convection. They concluded that the core/mantle boundary temperature is 3700 ± 500°C, which is within the range quoted above. They also found that the heat flow out of an adiabatic outer core must be (very approximately) 8×10^{12} W, which is close to the 10^{13} W postulated by Gubbins (1977). (These values of temperature and heat flow from the core are summarised, together with those of the mantle, in Table 8.2 and Figure 8.12.)

We shall now summarise present-day knowledge of the composition, state and evolution of the core. To have formed a core, the Earth must have lost oxygen and become chemically reduced relative to chondrites (see Eqs (5.1) and (5.2)), so that iron, along with other chalcophilic and siderophilic elements, was able to form a separate high-density zone at the centre of the Earth. It is agreed that much of the core formed early in the Earth's history, though there is some debate about just how long the process continued (Vollmer 1977 and Vidal & Dosso 1978). Initially, the core was probably a totally molten mixture, principally of iron, nickel and sulphur. If the Earth is chondritic in nickel (about 1%) and sulphur (about 8%), then the core contains about 3% nickel and 24% sulphur on the basis of efficient partitioning into the core. But the data of Ahrens (1979) indicate that the outer core contains only 9–12% sulphur, half that available in a C1 chondritic Earth. The remainder is in the crust and mantle or was lost during accretion. So a possible bulk composition of the core at formation was 86% iron, 11% sulphur and 3% nickel, with traces of other siderophilic and chalcophilic elements, possibly including up to 0.1% potassium. It is worth noting that most iron meteorites contain rather less sulphur than this, usually about 4–5%; nevertheless, troilite is an important mineral in these meteorites. (The latter had very short high-temperature histories and may not have had time for efficient core/mantle separation to have occurred.)

If the Earth's core began as a molten Fe–Ni–S system, then its temperature may have exceeded 4000°C (Fig. 6.7). However, it has cooled due to the steady loss of energy, mainly by thermal conduction to the mantle. At some stage, it must have cooled faster than it was heated by internal sources because the inner core has formed by crystallisation of a Fe–Ni alloy. Opinions differ about whether the core

105

is still cooling significantly in order for the inner core to continue growing and for gravitational convection to occur, or whether the inner core is almost stable in size with radioactive thermal convection predominating in the outer core. The inner core now comprises 1.7% of the Earth's mass and, if it contains about 20% nickel, this leaves the present-day outer core composition at about 86% iron, 12% sulphur and 2% nickel. Although this composition, 33% FeS, is close to the eutectic (about 48% FeS; Fig. 6.6b), further crystallisation can occur before the eutectic is reached and the core/mantle boundary temperature is still likely to be considerably above that of the eutectic (the best estimate for 33% FeS from Figure 6.6b is about 3200°C at the core/mantle boundary). This means that further growth of the inner core may be anticipated until it reaches about 10% of the mass of the Earth (about one third of the mass of the core). At this point, much of the inner core will be almost pure iron with all the available nickel, leaving a true eutectic mixture in the outer core which will be crystallising to form a solid core at about 1800°C.

Summary

1   From density data, shock compression studies and meteorite evidence, the Earth's core comprises:

  (a) an inner solid region, 1·7% of the Earth's mass, composed of iron–nickel alloy (probably roughly 20% Ni, 80% Fe), and
  (b) an outer fluid region, about 30% of the Earth's mass, composed of an iron–sulphur mixture containing roughly 12% sulphur and probably about 2% nickel.

2   The Earth's magnetic field can be used as a constraint on conditions in the core because its secular variation indicates a dynamic source. The field is thought to be the product of electrical current loops in the mobile outer core which, in the presence of their own associated magnetic field, perpetuate the field by dynamo action (the self-exciting dynamo). However, this requires that the outer core is convecting by thermal or other means.

3   The entire core is thought to have been molten in the Earth's early history, but to have cooled and crystallised the inner core. If this has been a progressive process, it may cause gravitational convection in the outer core due to the large density contrast between newly crystallised solid grains of metal and buoyant sulphide-bearing liquid. In this way, enough energy is available to generate the Earth's magnetic field by dynamo action ($10^9 - 10^{11}$ W are needed). Alternatively, the dynamo may be powered by thermal convection due to radioactive decay of ^{40}K in the core, which may be present because of the possible chalcophilic behaviour of alkali elements at high pressure. Precession and thermal convection due purely to crystallisation (latent heat release) are inadequate energy sources.

4   The results of melting experiments extrapolated to core pressures show that the Fe–Ni alloy of the inner core is solid because it melts at about 4500°C (inner core/outer core boundary melting temperature), whereas the outer core is fluid

because sulphur depresses the melting temperature of iron, forming an 1800°C eutectic at 48% FeS, 52% Fe (at the core/mantle boundary). As the outer core is more metallic than the eutectic, the actual temperature of the core lies between these extremes (i.e. in the range 1800–3900°C at the core/mantle boundary, and, most likely, at about 3200°C). The outer core is likely to have an approximately adiabatic thermal gradient and to deliver in the order of 10^{13} W to the mantle.

Further reading

General journal:
    Carrigan & Gubbins (1979): review of the source of the magnetic field.
General books:
    McElhinny (1973): systematic studies of the Earth's magnetic field and its origin.
    Bullard (1971): dynamo theory of the Earth's magnetic field.
Advanced journals:
    Lewis (1971); Ringwood (1977b); Loper (1978): the debate about radioactive and
        gravitational energy in the core.
Advanced books:
    Jacobs (1975): properties and constitution of the Earth's core; extensive bibliography.
    Ringwood (1975): core formation in relation to planetary accretion.
    Garland (1971); Stacey (1977): dynamo theory of the Earth's magnetic field.

7 The mantle and oceanic crust

7.1 Introduction

The mantle can be subdivided into three main seismic regions which are broadly concentric with the Earth's surface: the upper mantle, the Transition Zone, with anomalous velocity gradients, and the lower mantle. Values of density within these regions are shown in Figure 3.14. Rather low density gradients occur in the top 400 km, and in the Transition Zone there are steep increases of density with depth; below 1050 km, extending through a broad zone down to the core/mantle boundary at 2885 km depth, there are low density gradients, with the possible exception of another transition region right at the base of the mantle. The mineralogical and chemical composition of the mantle is much less well known than the density, but the available information comes from physical constraints, meteorite data and geological materials.

(a) *Physical constraints* are used to determine possible rock types that may exist at depth. In addition to density the closely related property of **lithostatic pressure** (i.e. that due to the weight of overlying rocks) must be taken into account. *Temperature* is also important for various reasons related to melting and convection (see Sections 7.2 and 8.6). Given the pressure and temperature inside the mantle, as well as related values of density, physical measurements on supposed mantle rock samples can be made to determine how representative those rocks may be. In the following text, there are references to possible deep mantle compositions whose densities are referred to surface pressures – **'zero' pressure densities**.

(b) *Meteorite data* provide a check on possible compositions. Bearing in mind the Chondritic Earth Model (Section 5.4), the Earth's original mantle should have resembled the silicate phases of chondrites. Together with the theory of element distributions according to their electronic properties (Section 5.3), this evidence gives a further constraint on bulk composition and composition–depth variations.

(c) *Geological materials* known to come from the mantle are vital. These include the melting products of the mantle seen at the surface – volcanic *basalts* – and the *nodules*, or fragments, of supposed mantle material they contain. A link between possible mantle compositions and their melting products is established through **experimental petrology** which allows temperatures and pressures to be reproduced for at least the top 600 km or so of the mantle. The source depth of basalts is well within this depth range, as is shown by the earthquakes related to volcanic eruptions (Eaton & Murata 1960).

Other useful geological materials are those from **kimberlite pipes**, which have sources at mantle depths, and **ophiolites**, which comprise both ocean crust and upper mantle rocks that have been upthrust by Earth movements. Ophiolites have such an important part to play that it is quite impossible to consider the composition of the mantle in isolation from the oceanic crust. There is a great deal more information about the upper layers of the Earth, by virtue of their accessibility to sampling and to experiment, than about the lower parts of the mantle. In this chapter, we consider the three density zones outlined earlier, as a prelude to describing the dynamic nature of the crust and mantle, together with their evolution, in Chapters 8–10.

7.2   The upper mantle: eclogite or peridotite?

7.2.1   *A preliminary model*

One approach to mantle composition is to ask what materials could be the source of the basalts that form almost the whole of the oceanic crust and that are also quite common on land. The search for such rocks is easily narrowed down to two categories: peridotite or eclogite.

(a) *Peridotite* – a group name given to an extensive category of ultrabasic rocks, typically composed of about 80% olivine and 20% pyroxene. Peridotites occur as wedges in certain young fold mountain belts, on certain oceanic islands (mainly as nodules in basalt) and in the diamond-bearing kimberlite pipes of ancient continental regions, e.g. South Africa and Western Australia (see Fig. 7.1). Kimberlite pipes, formed by explosive solid–gas volcanism, contain fragments of garnet-rich peridotite, some eclogite (see below) and often diamonds, all in a finer-grained matrix dominated by micaceous minerals.

Figure 7.1   A kimberlite pipe from which the garnet peridotite has been mined for diamonds. (From an original picture by P.G. Harris.)

109

(b) *Eclogite* – a high-pressure, low-temperature metamorphic rock whose bulk composition closely resembles basalt. Mineralogically, eclogites contain roughly equal parts of aluminous pyroxene and the dense mineral, garnet; they also occur in young fold mountain belts, such as the Alps and Himalayas, and are thought to be metamorphosed basalts:

plagioclase feldspar + pyroxene + olivine → garnet + aluminous pyroxene + quartz
basalt   **eclogite**   (7.1)

The basic differences between eclogite and peridotite are that the former contains more garnet whereas the latter is dominated by olivine; eclogite also contains more pyroxene and is more silica-rich than peridotite.

It is useful to consider the nature of the crust/mantle boundary, or Moho seismic discontinuity, in each case. Above this boundary, the ocean crust is basaltic in composition (Section 7.3), whereas the continental crust is chemically and mineralogically rather different (tonalites and granulites predominate – Chs 9 and 10). From Equation (7.1), if the upper mantle is eclogitic, then the oceanic Moho represents a **phase change** from a low- to a high-pressure form of the same basaltic composition. In contrast, for a peridotite upper mantle, the oceanic Moho would represent a **compositional change** from a basaltic, basic composition crust to a peridotite, ultrabasic upper mantle. In both cases, the *continental* Moho would represent a compositional change.

By subjecting samples to appropriate pressure–temperature conditions it is found that a phase change conflicts with the observed depth to the oceanic Moho. To appreciate the experimental evidence, note that a phase change would occur at a given pressure, and hence a constant depth, were it not for differences in temperature gradient (Fig. 7.2). Higher temperatures tend to cause expansion and so favour lower-density basalt, whereas lower temperatures would favour higher-density eclogite if suitable pressures exist. Just what are the pressure–temperature conditions at the Moho and in the upper mantle?

Pressure (P) varies according to depth (h) and the density of overlying material. If, for simplicity, we assume an average density of $3300 \, \text{kg m}^{-3}$ in the top few hundred kilometres of the Earth, we have

$$P = 3 \cdot 3 \times 10^7 h \quad \text{N m}^{-2}$$

where h is in kilometres. Or, to use units which are geologically more appropriate, we can write

$$P = 0 \cdot 33 h \text{ kilobars (kbar)}$$

Temperature gradients are a product of several factors, such as the heat productivity of the rocks, their thermal conductivities and the type of heat transport: convection or conduction. Near-surface crustal rocks are relatively rigid compared with deeper mantle materials, so that they resist heat transport by convection. They also contain more radioactive heat sources than any postulated mantle rocks and so the crust is characterised by the highest thermal gradients (Chs 8 and 9). Using boreholes, measured thermal gradients in shallow rocks are found to lie between 20 and $40°C \, \text{km}^{-1}$ but such gradients cannot be extrapolated through the entire 2900 km of mantle. Apart from the lower heat productivity of mantle rocks and their plastic behaviour (Ch. 8) we know that a reasonable upper limit for the outer core temperature is about 4000°C (Section 6.3). At depth, the mantle thermal gradients must decrease towards the adiabatic gradient (Section 3.4), roughly $0.3°C\,km^{-1}$. Such curved geotherms (Fig. 7.2) comprise the steeper gradients of a rigid, grossly superadiabatic conducting layer (the crust and uppermost mantle) and the shallower gradients of a mildly superadiabatic convective layer below.

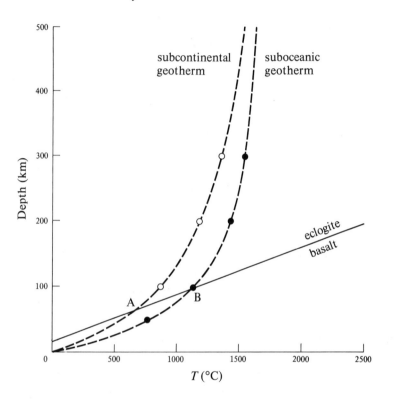

Figure 7.2 Estimated geothermal gradients in the upper mantle and their relationship to the basalt/eclogite phase boundary. If the upper mantle were eclogitic, then the subcontinental Moho would be at (A) and the suboceanic Moho at (B). (After Ito and Kennedy 1970.)

Apart from these gross features, there are also minor differences between the thermal gradients of near-surface oceanic and continental regions (Fig. 7.2). This is related to the near-equivalence of equilibrium oceanic and continental heat flow (Ch. 8). Because observed continental rocks have much higher heat productivity than most oceanic rocks, it is postulated that suboceanic temperatures in the upper mantle must be higher than in the subcontinental mantle.

Now we can return to the experimental evidence for the basalt–eclogite phase transition. The precise location of this boundary is not known with confidence, but the higher the temperature at depth, the deeper must be the phase change. So, in regions of high geothermal gradient, the Moho for an eclogite upper mantle would be *deeper* (point B in Fig. 7.2) than in regions of lower geothermal gradient such as the continents (point A). No such correlation between geothermal gradient and depth to the seismic Moho is found and, furthermore, it is known that higher pressures than those at the oceanic Moho are required for the production of eclogite. One final nail in the coffin of the eclogite model is that 100% melting would be required to produce chemically identical basalt magma. A fully liquid layer

111

would completely attenuate S-waves, which does not occur. The small degree of attenuation observed is compatible with only a few percent partial melting, and so favours the *peridotite* model. Thus, the fact that eclogite is compositionally very similar to basalt, far from making it a likely source material, rules it out completely!

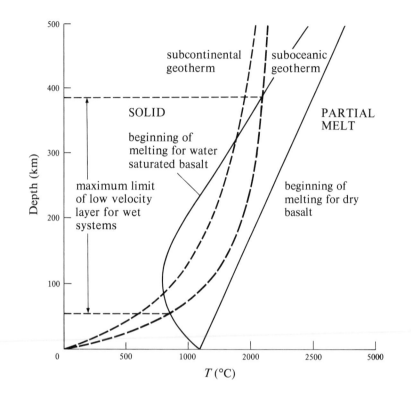

Figure 7.3 Upper mantle geothermal gradients (dashed, from Fig. 7.2) and their relationship to wet and dry melting curves for peridotite (solid) to give basalt partial melt. If excess water were present, the low-velocity layer would reach its maximum theoretical limits between the horizontal dashed lines.

7.2.2 *Basalt melting: the experimental evidence*

Experimental evidence shows that *partial* melting of peridotite gives basalt at upper mantle conditions (Fig. 7.3) such as we believe to occur in the low-velocity zone. The amount, or degree, of partial melting is discussed more fully in later sections and here we consider the *conditions* of melting using the same thermal gradient curves as in Figure 7.2. The first basaltic liquids in a *dry* system are produced along a curve of positive pressure–temperature slope starting at 1200°C and 1 atm pressure (Fig. 7.3) – to the left is solid peridotite and, to the right, partial melt plus residue. As the field of dry basalt partial melt does not intersect the geothermal gradient curves (Fig. 7.3) no melting could occur were it not for two additional factors:

(a) Exceptional circumstances may exist where the geothermal gradient is perturbed enough to reach the dry melting curve for basalt. Hot spots in the mantle have been postulated in various places; for example, to account for oceanic islands such as the Hawaiian chain (see Section 8.7). This factor could be particularly important where high thermal gradients are localised beneath ocean ridges. Quite apart from the higher temperatures of such regions, melting may also be favoured by the tendency of ocean lithosphere to spread away from such zones (Section 8.4). As new material rises, pressure is lowered rapidly but temperatures remain almost constant and so, from Figure 7.3, pressure release may aid melting.

(b) An equally important local control is the presence and concentration of volatiles such as carbon dioxide and water. For example, if uncombined water is present, then the melting temperature may be lowered by as much as 500°C. The curve for water-saturated melting has a negative pressure–temperature slope at low pressures, but becomes positive with increasing depth (Fig. 7.3). Different amounts of water are required to saturate a melt at different pressures, but an average amount would be 10%. Such large quantities are not available in the mantle, but small amounts (roughly 0·1%) would still allow significant partial melting to occur. Although melting will *start* at the water-saturated curve, only very limited amounts of melt can be produced here, even though all the water available concentrates in the melt. But the water requirement *decreases* with increasing temperature so that more melting could occur with the same water content.

Therefore, water has two important effects; first, it lowers melting temperatures so that fusion can occur where it would not be possible in a dry system and, secondly, the amount of water may be more important than lithostatic pressure and temperature in controlling the *amount* of melt produced.

At first sight, it may seem unlikely that water could be present in any form deep in the mantle, but there exist several mineral groups that contain crystalline water which is locked up in their silicate structures as hydroxyl groups. Provided that traces of these minerals are present they will decompose, releasing water and causing melting. Two minerals are thought to be significant:

(a) *Hornblende*, an amphibole, whose simplified formula is $Ca_2(Mg, Fe)_5(Si_4O_{11})$ $(OH)_2$ and which releases its water in the range 700–1100°C.
(b) *Phlogopite*, a mica, $KMg_3(AlSi_3O_{10})(OH)_2$ which is stable up to the range 900–1300°C.

The temperature range (700–1300°C) correlates with the depth range of peridotite melting between maximum limits of 50 and 400 km (Fig. 7.3), thus defining the extreme limits of the low-velocity zone in a normal oceanic environment. Melting is more usually confined to the upper part of the low-velocity zone (50–200 km) where there is a greater difference between the geothermal gradient and wet melting curves. It should be noted also that other volatile components, such as CO_2, may exert controls similar to water on the conditions of basalt melting.

Finally, not all basalts found in oceanic environments have the same composition and mineralogy. Most fall into one of two categories: **tholeiites**, which are relatively rich in silica and poor in potassium and other alkali elements (roughly 50% SiO_2 and 0·5% K_2O), and **alkali basalts**, which are lower in silica but contain more alkalis (roughly 46% SiO_2 and 1–2% K_2O). Tholeiitic basalts are usually found on ocean ridges and other regions where thermal gradients are exceptionally high (locally, about 100°C km^{-1}) and they originate by partial melting at shallow depths. Alkali basalts are more typical of ocean islands and originate from greater depths (usually 50–100 km) in regions of lower thermal gradient (< 30°C km^{-1}) than

113

tholeiites. They may also require lower degrees of partial melting than tholeiites to concentrate the scarce alkali elements.

Yet a third type of basalt occurs in island arcs and Andean-type continental margins. These are the **calc-alkaline basalts**, sometimes termed high-alumina basalts because of their higher concentrations of Al_2O_3 (about 17%) than in other basalts (11–16%). In such locations, calc-alkaline basalts are usually subordinate in volume to **andesites** (intermediate lavas averaging 55% SiO_2 which are also calc-alkaline). The origin of these magmas is related to the downwards return flow of oceanic lithosphere into the deep mantle beneath oceanic and continental arcs (see Chs 9 and 10).

7.3   The upper mantle and oceanic lithosphere: a detailed view

7.3.1   *The evidence of direct samples*

A great advance in our understanding of the mantle came with the advent of Plate Tectonic theory (see Section 8.4). It was recognised that ocean floors are transient features, deriving from the mantle and returning to it in a time of the order of 100 Ma. This has many implications and of immediate concern is that occasional portions of oceanic plate escape destruction at subduction zones in the final stages of ocean closure when part of the ocean plate is 'pinched up' and carried or **obducted** onto one of the forelands of two colliding continents, to give rise to a characteristic **'ophiolite suite'**. Many ophiolite sequences have been recognised (Coleman 1977 and Gass & Smewing 1980 give reviews) and perhaps the best known occurs in the Troodos Massif of Cyprus. The obduction of Troodos occurred during the closure of the Tethyian (Mediterranean) Ocean during Tertiary times, and the actual underthrusting of Cyprus by the African continental foreland was a Miocene event.

The significance of ophiolite studies is that the petrology of ocean crust and upper mantle layers can be related to the more general seismic properties of these layers which characterise all oceanic regions (Fig. 7.4). Of course, the location of ophiolite

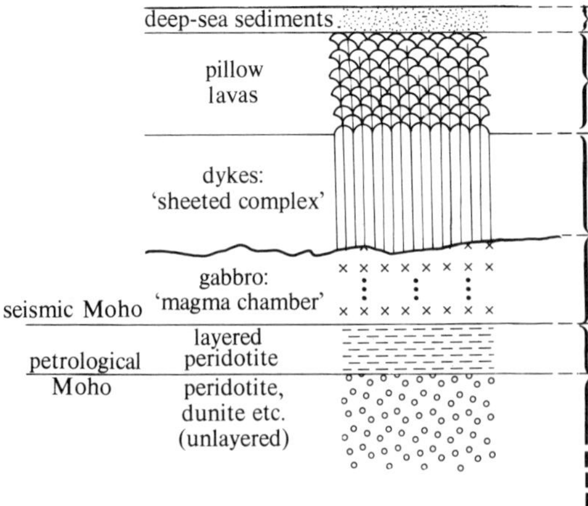

Figure 7.4 Petrological, seismic and thickness data for a typical ophiolite sequence compared with seismic layers for the ocean crust; the layers are not drawn to scale. (Data from Cann 1970, Matthews *et al.* 1971, Christensen & Salisbury 1975, Gass & Smewing 1980.)

		Typical ophiolite		Normal ocean crust	
		thickness (km)	P-wave vel. (km s^{-1})	thickness (km)	P-wave vel. (km s^{-1})
deep-sea sediments	layer 1	*c.* 0·3	–	0·5	2·0
pillow lavas	layer 2	0·3–0·7	3·3	1·7	5·0
dykes: 'sheeted complex'		1·0–1·5	4·1	1·8	6·7
gabbro: 'magma chamber'	layer 3	2–5	5·1	3·0	7·1
layered peridotite					
peridotite, dunite etc. (unlayered)	layer 4	up to 7·0	2·5–4·2	–	8·1

seismic Moho

petrological Moho

uncertainty:

(a) the ophiolite sequence is usually rather thin, containing 2–4 km of ocean crust
materials overlying mantle, compared with 5 km of crust in most parts of the
oceans;
(b) none of the magnetic anomalies that characterise ocean crust elsewhere have yet
been discovered from ophiolites; and
(c) seismic velocities are generally much lower in ophiolites than in normal ocean
crust (Fig. 7.4).

There are various explanations of these uncertainties. For example, ophiolites are
often 'dismembered' by the thrusting process, but the thin sequences are more often
attributed to emplacement, soon after formation, from back-arc marginal basins, as
in the modern western Pacific, or from zones of ocean floor close to the spreading
ridge zone. Magnetic anomalies may be lost through extensive hydrothermal
metamorphism, usually observed in ophiolites, although most of the ophiolites so
far studied comprise ocean lithosphere formed during a long period of normal
polarity during late Mesozoic times. So far as seismic velocities are concerned, again
hydrothermal metamorphism has caused extensive mineralogical changes into
expanded clay mineral and serpentine lattices which replace all the primary silicate
minerals, particularly olivine. These changes, along with the cracking and fissuring
of uplifted materials, all act to reduce seismic velocities. Despite the differences
between ophiolite and normal ocean crust data on the right of Figure 7.4, it seems
that there are strong similarities between the two.

Accepting that ophiolites really do represent ocean lithosphere, what information
do they provide about the materials present and their evolution? The upper portion
of the sequence comprises the fine-grained Fe–Mn-rich mudstones, cherts, shales
and limestones of deep-sea sediment cores. Both these sediments and the underlying
tholeiitic *pillow lava* sequence contain heavy-metal (Cu–Zn) sulphide deposits,
apparently deposited by hydrothermal mineralising solutions when high thermal
gradients operated across the newly formed ocean lithosphere. Pillow lavas (Fig.
7.5) are a product of chilling by cold sea water and were apparently fed by a series of
multiple dykes – the so-called sheeted complex – from a basic magma chamber, now
in the form of coarse-grained *gabbros* (Fig. 7.4). Figure 7.5 illustrates the lower part
of layer 2 with pillow lavas being cut by later dykes. A remarkable feature of these
dykes is that they have chilled margins on one side only, due to their origin as
magma sheets which intrude ocean ridge zones and cool preferentially on one side
as they spread away. Continuing downwards, the dykes are replaced by plutonic
gabbros at the base of seismic layer 3. These intrusive rocks are notable for their
crystal fractionation textures: early-crystallising minerals were mainly olivine with
some pyroxene. Both are relatively dense compared with basic magma and so sank
to the base of the chamber and formed **crystal cumulates** – the layered peridotites –
at the top of seismic layer 4 (Fig. 7.4). (Small volumes of low-alkali silicic rocks
(plagiogranites) sometimes form from the final **residual liquids** that crystallise near
the top of basic magma chambers.) The seismic distinction between ocean crust and 115

upper mantle is thought to occur where gabbros grade downwards into layered peridotites at the layer 3/layer 4 junction in Figure 7.4. But the petrological boundary occurs where layered peridotites, produced by gravity settling in *crustal* magma chambers, change to the massive peridotites that have always been part of the upper mantle.

Figure 7.5 Pillow lava/dyke intersection in Cyprus ophiolite. The dykes are the vertical walls of rock on the right and the pillow lavas are horizontal. (From an original photograph by F. J. Vine.)

There are still the following questions: (i) How varied is *peridotite* in ophiolite sequences? And (ii) how typical are ophiolitic rocks of the remainder of the mantle? We need to distinguish undepleted mantle material from that which has been depleted of certain chemical components through magma extraction. The un-layered, massive ultrabasic bodies of ophiolite sequences are often chemically complex and are strongly metamorphosed with gneissose textures. The prominent rock types are:

(a) *harzburgite*: mainly olivine (about 80%) with minor orthopyroxene (about 20%);
(b) *dunite*: almost entirely olivine; and lesser amounts of
(c) *lherzolite*: again dominated by olivine (about 60%), but with orthopyroxene and clinopyroxene (about 30%). Other minor components include spinel ($MgAl_2O_4$), garnet ($CaMg_2Al_2Si_3O_{12}$) and plagioclase ($CaAl_2Si_2O_8$). Plagioclase is a common minor component of ophiolitic lherzolites.

To these, we now add the evidence of other materials from the upper mantle:

(a) Peridotites, found as nodules in kimberlite pipes (Fig. 7.1). Some kimberlites come from more than 150 km depth, because the diamonds that they sometimes contain could only be formed at such depths. (Diamond is a high-pressure phase of carbon, or graphite.) Their most common rock type is garnet lherzolite with smaller amounts of spinel lherzolite, garnet harzburgite and harzburgite. Evidently, garnet and pyroxene are much more important constituents at these depths (about 100–200 km) than in the top 5 km of the mantle, represented by ophiolites, but olivine still predominates.

(b)   Ultrabasic nodules, or xenoliths, which occur in the basalts of many oceanic islands and of some continental volcanoes (Fig. 7.6). These nodules are usually 1–10 cm in size and most are thought to be *cognate*, i.e. they are pieces of depleted residuum from partial melting brought up from the basalt melt source region. Because basalts come from a range of depths, so too do these ultra-basic nodules, but their occurrence with basalts – mantle partial melts – means that the nodules cannot represent fresh, undepleted mantle. Most of them are rich in olivine which has the highest melting temperature and is, therefore, the most refractory of the major mantle minerals. Other interesting compositions occur less frequently. These include different kinds of lherzolites, mainly spinel-bearing, from the alkali basalts of oceanic islands which were melted from regions in the 50–100 km depth range (Section 7.2).

Figure 7.6 Olivine nodule in basalt lava. The outer part of this sample is dark grey, fine-grained basalt. In the centre is a coarse aggregate of green olivine crystals.

How are these different upper mantle samples related to each other and to partial melting processes? One way of establishing this relationship is to compare the principal mantle rock types chemically in order to distinguish residual and relatively pristine material. The four analyses shown in Table 7.1 are averages and form the basis only of a rough comparison between the groups. Of the three possible mantle compositions garnet lherzolite is closest to the composition of basalt and so it is the *least* depleted. For example, elements that show the greatest variation across the table: TiO_2, Al_2O_3, CaO, Na_2O and K_2O all enter the melt phase in preference to the residuum. (A rigorous explanation of the partition of elements between melt and residuum is beyond the scope of this book, but appears in specialist geochemistry texts.)

Table 7.1 Typical chemical compositions (percentages) of principal upper mantle rock types and oceanic basalts.

	1 Ophiolitic harzburgite	*2* Olivine nodules	*3* Garnet lherzolite	*4* Oceanic basalt
SiO_2	42·3	44·5	45·3	47·1
TiO_2	0·1	0·1	0·2	2·3
Al_2O_3	0·5	1·7	3·6	14·2
FeO	7·1	9·6	7·3	11·0
MnO	0·1	0·1	0·1	0·2
MgO	49·6	42·3	41·3	12·7
CaO	0·1	1·6	1·9	9·9
Na_2O	0·1	0·1	0·2	2·2
K_2O	0·005	0·04	0·1	0·4

Figure 7.7 focuses on garnet lherzolite as a possible undepleted mantle source rock from which oceanic basalts could be melted, following which, the composition of the residuum would be driven back in the opposite direction. The concentrations of the two elements plotted are almost negligible in harzburgite and so we might regard this as the most depleted residual mantle material yet sampled – even more so than the olivine nodules, which are partially depleted. If all the available K_2O of garnet lherzolite (0·1%) were concentrated into the basalt melt (0·4% K_2O), effectively leaving none in the residuum (harzburgite has only 0·005% K_2O), then a concentration factor of 4 times is involved. This places an absolute *maximum* limit of 25% on the degree of partial melting involved to produce the basalt in Table 7.1. Whereas some of the K_2O of garnet lherzolite might easily concentrate into lesser volumes of melt, leaving a source only partially depleted in potassium, more than 25% partial melting of garnet lherzolite could not yield a melt product so rich in K_2O as oceanic basalt. The same sort of exercise can be applied to other chemical and mineralogical phases and with similar conclusions. But this does not mean that the mantle has everywhere melted to this degree, for various reasons. First, alkali basalts contain more K_2O than tholeiites and, moreover, come from sources that tap the normal oceanic low-velocity zone where independent evidence indicates an average of less than 5% partial melting (Section 8.5). Secondly, the garnet lherzolite analysis quoted in Table 7.1 may not be typical of all regions, since it is based on continental kimberlite nodules: the mantle may have less than 0·1% K_2O on average. Thirdly, materials so depleted in K_2O, etc., as the harzburgite analysis in Table 7.1, are only known from the *topmost* mantle, as seen in ophiolite complexes where tholeiitic ocean crust has been extracted at high thermal gradients. Only in these upper zones of the mantle may the maximum limit of melting have been reached.

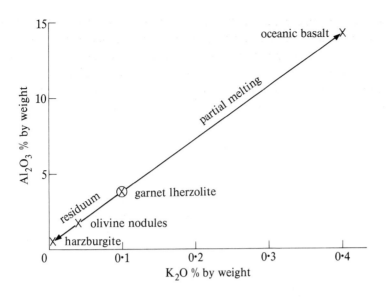

Figure 7.7 Plot of Al_2O_3 versus K_2O for mantle rock types and oceanic basalt in Table 7.1, also showing the directions in which partial melt and residuum progress from a garnet lherzolite parent.

Figure 7.8 is a schematic illustration of a mid-ocean ridge where new mantle-derived melt is continually added to the crustal magma chamber. The two sides spread apart and ocean crust crystallises, but in the topmost mantle the material that spreads away has been depleted by melting under the ridge. A two-stage process is represented. First, the mantle partially melts and the magma rises to occupy a convecting magma chamber in the ocean crust. Secondly, crystallisation of high-temperature minerals in the magma chamber causes the layered peridotites to be precipitated, and this is accompanied by occasional ejection of magma to the

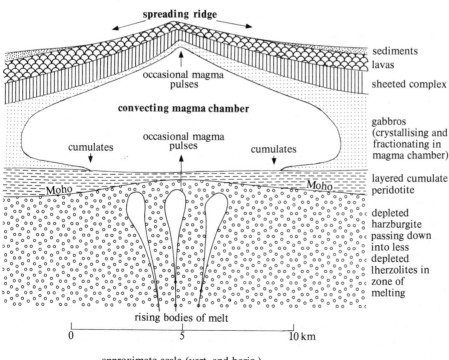

Figure 7.8 Schematic section through a mid-ocean ridge showing the evolution of ocean crust/upper mantle structure and composition. See text for further explanation. (Based on Cann 1974, Gass & Smewing 1980, with simplifications.)

119

surface until, eventually, the remaining magma crystallises as coarse-grained gabbro. The upper mantle residuum is the harzburgite and dunite which are found in the massive peridotites of ophiolite sequences. Patches of undepleted lherzolite which *are* found in ophiolites must represent zones where partial melting has not been effective and this is usually plagioclase lherzolite (Section 7.3.2). Beneath the topmost oceanic mantle, sampled by ophiolites, the peridotites of olivine nodules and kimberlites have apparently suffered less effective depletion processes.

Here is a summary of the arguments so far. The formation of 5 km of ocean crust at the expense of 20–30 km of the topmost oceanic mantle by up to 25% melting has depleted this zone of its low-temperature-melting components. The evidence from olivine nodules and kimberlites indicates that the mantle becomes less depleted with increasing depth, and that a close approach to pristine mantle may be provided by lherzolite. At low pressures, samples of plagioclase lherzolite are known from ophiolites; in the zone of alkali basalt magma generation (50–100 km depth), the material seen in occasional undepleted nodules is spinel lherzolite, and in kimberlite pipes (100–200 km source depth) it is garnet lherzolite. Chemically, these lherzolites are very similar, though their densities differ (see below). Traces of hydrated minerals, necessary for wet melting are also occasionally found in lherzolites. Kimberlite pipes, in particular, are noted for a prominent mica component, though some of this may be of low-pressure, secondary origin.

7.3.2 *Pyrolite and lherzolite: an upper mantle model*

A completely separate approach to determining the composition of the upper mantle by modelling was developed before the widespread recognition of the importance of lherzolites. The **pyrolite model** was proposed by Green and Ringwood in 1963 who also recognised that most of the available mantle samples, dunite, harzburgite etc. had probably undergone partial melting already to produce basalt. Therefore, they suggested that a model composition for the undepleted upper mantle could be subjected to experimental melting to determine the types of basalt melt that it produced. Their mixture comprised three parts of the most depleted material available, i.e. dunite, and one part tholeiitic basalt. This they called **pyrolite** (pyroxene–olivine rock). In Table 7.2 their chemical composition for pyrolite is compared with the chemistry of the silicate phases of chondritic meteorites and the garnet lherzolite analysis from Table 7.1. Pyrolite was made to differ from chondritic silicates in two ways:

(a) chondrites have an Fe/Mg ratio of 0·45, whereas that of pyrolite was made 0·2 to be consistent with the available evidence from nodules and peridotite samples; and

(b) in the mixture of dunite and basalt which comprises pyrolite, the silica concentration is lower than in chondritic silicates, giving a lower pyroxene/olivine ratio.

At the time that the pyrolite model was conceived, Ringwood was postulating that all the excess *iron and silicon* of a chondritic Earth had been separated into the core. Nowadays, it is much less certain, on chemical equilibrium grounds, whether

Table 7.2 Chemical compositions of chondritic silicates, pyrolite and garnet lherzolite.

	1 *Chondritic* *silicates*	*2* *Pyrolite*	*3* *Garnet* *lherzolite*
SiO_2	48·1	45·2	45·3
TiO_2	0·4	0·7	0·2
Al_2O_3	3·8	3·5	3·6
FeO	13·5	9·2	7·3
MnO	0·2	0·14	0·1
MgO	30·5	37·5	41·3
CaO	2·4	3·1	1·9
Na_2O	0·9	0·6	0·2
K_2O	0·2	0·13	0·1

silicon occurs as a core diluent (Ch. 6). Point (b) above would probably not be accepted in a modern version of pyrolite, but the difference in SiO_2 concentration between chondrites and pyrolite makes little difference to the results of experimental studies. Apart from this and the Fe/Mg ratio, the most striking feature of Table 7.2 is the similarity of pyrolite and garnet lherzolite, two independent estimates of the upper mantle composition, to chondritic silicates. Experimental studies on pyrolite at values of pressure, temperature and volatile content applicable to the upper mantle have yielded the entire range of melt products, ranging from alkali basalts to tholeiites to rare varieties such as picritic, magnesium-rich lavas, as partial melting proceeds.

Given that pristine upper mantle is made of something chemically like pyrolite or garnet lherzolite, what variations in mineral assemblage should we expect, apart from those due to melting?

Innumerable variations are known and the most important is summarised by the following reaction:

$$Mg_2SiO_4 \;+\; CaAl_2Si_2O_8 \longrightarrow CaMg_2Al_2Si_3O_{12} \qquad (7.2)$$
$$\text{olivine} \qquad \text{feldspar} \qquad\qquad \text{garnet}$$

This reaction in basic and ultrabasic rocks, which replaces a little of the olivine and all of the feldspar by garnet, proceeds to the right with increasing pressure because garnet is a relatively dense mineral phase. However, in 1966, Kushiro and Yoder found that there is an intermediate step in the reaction where spinel ($MgAl_2O_4$) becomes stable together with aluminium-bearing pyroxenes. The two reactions are:

(i) olivine + feldspar → clinopyroxene + orthopyroxene + spinel

(7.3)

(ii) clinopyroxene + orthopyroxene + spinel + more feldspar → garnet

The relationship of these assemblages to pressure and temperature is shown in Figure 7.9. The phase boundaries shown are considerably blurred in natural systems, particularly due to minor elements in some of the minerals. A particular example is the development of chrome spinel ($FeCr_2O_4$) which, even in small quantities, tends to move the spinel–garnet transition to pressures several kilobars higher than for pure $MgAl_2O_4$. Figure 7.9 provides the basis for understanding why shallow (ophiolitic) samples of pristine mantle are plagioclase lherzolite, giving way

121

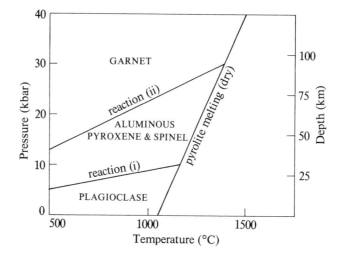

Figure 7.9 Relationship of plagioclase, spinel and garnet stability in lherzolite relative to the pyrolite melting curve in the upper mantle. The reactions mentioned are those in Eq. 7.3.

downwards to spinel lherzolite (some nodules in alkali basalts) and eventually garnet lherzolite (from kimberlite nodules).

A schematic summary of the composition of the upper mantle is given in Figure 7.10. Further details of the subcontinental mantle appear in Section 8.6 and Chapter 9 but here are a few preliminary comments. Seismic data show that the continental crust is about four times as thick as the ocean crust (Section 2.3) and there is increasing evidence that ocean–continent differences extend to considerable depths, perhaps 700 km, into the mantle. The subcontinental mantle has the higher P-wave seismic velocities, probably indicating that it represents a deep, refractory root zone from which the continental crust was derived. Together, the crust and upper mantle comprise a thick, rigid mobile plate. This is why a large zone of depleted mantle peridotite appears beneath the continent in Figure 7.10.

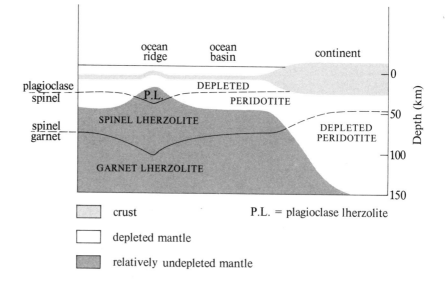

Figure 7.10 Schematic cross section through the top 150 km of mantle beneath oceanic and continental regions, ignoring the complications of a subduction zone. The junction between depleted and undepleted mantle in *oceanic* regions corresponds to the top of the low-velocity zone. Stability regions for plagioclase, spinel and garnet lherzolite are based on Figure 7.9 using appropriate thermal gradients (see Fig. 7.2 and Section 7.2).

Beneath the oceans, the depleted mantle thickens rapidly away from the active ridge zone (Fig. 7.8) and, under the ocean basins, it comprises harzburgites and dunites together with slices of undepleted lherzolite. The base of the depleted mantle zone is thought to be approximately isothermal, corresponding closely to the top of the low-velocity zone – shallow beneath the ocean ridges (about 5–30 km) and deeper elsewhere (Fig. 7.3). Relatively undepleted lherzolite exists at greater depths with the stability boundaries extrapolated from Figure 7.9 to Figure 7.10. These boundaries bulge downwards into the mantle beneath the ridge zone because the reactions occur at higher pressures if higher temperatures apply (Fig. 7.9). They are also extrapolated into the depleted peridotite region of Figure 7.10 to indicate the nature of any remaining lherzolite slivers. In the low-velocity zone (50–200 km depth), the material comprises partially molten lherzolite, probably still in a fairly pristine state, because only small volumes of melt are extracted locally by volcanic activity.

The change from spinel to garnet lherzolite involves a small increase in density of about 60 kg m^{-3} in response to increasing pressure. Garnet lherzolite characterises the broad region from about 100 km depth down to the top of the Transition Zone at 400 km. Few volcanic liquids emanate from this depth and the only surface representatives are the kimberlite mixtures. It seems that the unique combination of high enough volatile vapour pressures and temperatures is rarely achieved at these depths.

Huge (sub-continental) thicknesses of depleted peridotite do not accumulate beneath the ocean crust because the oceanic lithosphere is resorbed into the mantle via subduction zones within 10^8 years or so of its formation. But at subduction zones it contributes to the formation of calc-alkaline magmas and, therefore, the resorbed material has a net composition more refractory and ultrabasic than before it produced oceanic lithosphere at the spreading ridge. This means that the mantle has progressively changed its entire (bulk) composition throughout the Earth's history and this may have led to large-scale chemical heterogeneities (Section 8.9 and Ch. 10).

7.4   The Transition Zone

Seismic studies of this more remote and inaccessible part of the mantle, between 400 and 1050 km depth, have defined three steps in the velocity–depth profile at about 400, 650 and 1050 km (Fig. 7.11). These steps correlate with jumps in the density–depth profile (Fig. 3.14) and a total increase in density from 3500 to 4600 kg m^{-3} occurs across the zone. In the 1930s, Birch and Bernal first proposed that the Transition Zone represents a region of *high-pressure transformations* for silicate minerals (Birch 1961). The new minerals, appropriate to the conditions of pressure and temperature in the lower mantle, would be denser and have more closely packed internal atomic structures. Recently, experimental phase changes in suitable materials have accounted for both the observed seismic velocity and density changes. A phase transformation involves changes in structure and density, but not in composition (Eq. (7.1) is an example). Another example is the transformation of crystalline carbon from graphite (density of about 2×10^3 kg m^{-3}) to diamond (density roughly $3 \cdot 5 \times 10^3$ kg m^{-3}) at upper mantle pressures. Graphite consists of carbon atoms in sheets with relatively large inter-sheet spacings, whereas diamond has short uniform bond lengths between all its atoms.

123

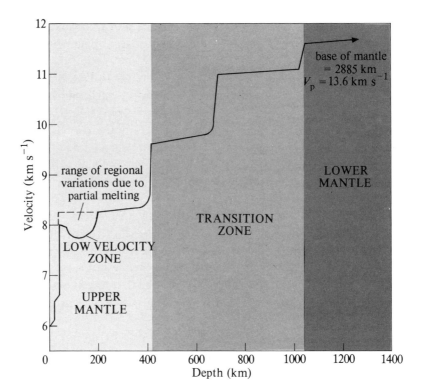

Figure 7.11 Simplified
illustration of seismic P-wave
velocity variations in the
upper mantle and Transition
Zone.

7.4.1 *The evidence for phase transformations*

The main difficulty in determining the phase changes at Transition Zone pressures has been technical: until the 1960s it was not possible to simulate the pressures and temperatures of mantle depths beyond about 200 km. But earlier workers had experimental data for **germanate** systems, which have identical mineral structures to silicates except that their phase transformations occur at much lower pressures. Germanate minerals are analogous to silicates because they are built from GeO_4^{4-} polytetrahedra, just like SiO_4^{4-} groups, except for a larger cation at the centre (Ge^{4+} radius = 53 pm; Si^{4+} radius = 42 pm). Using these analogues, it was discovered that magnesium germanate (Mg_2GeO_4) transforms from a relatively open ortho-rhombic structure to a more closely packed, cubic, *spinel-like structure*, with an 8% increase in zero-pressure density, at about 40 kbar pressure. This implies that natural olivine (Mg_2SiO_4) should undergo a similar transformation in the mantle.

In the late 1960s, a high-pressure apparatus was developed which reaches 200 kbar, equivalent to over 600 km depth. Earlier predictions based on germanate analogues were verified by synthesising the high-pressure silicate phases directly, but our knowledge of the region below 600 km rests largely on inference from germanates and experiments with explosive shock waves. Although the high-pressure shock waves, up to several megabars, are produced for only a microsecond, they allow certain physical properties of the high-pressure phases produced, such as density, to be measured instantaneously.

Most of the important phase change reactions in the mantle are **reconstructive transformations**, in which the high- and low-pressure polymorphs are structurally very different (graphite and diamond, for example). A large energy barrier must be

124

overcome to break bonds and reorganise the constituent atoms. Like diamonds, the high-pressure silicate polymorphs do not readily revert to their low pressure forms immediately after synthesis experiments and, so, their zero-pressure densities may be measured directly. However, because of their short duration, shock-wave experiments do not run to equilibrium, their products are unstable and high-pressure densities must be measured as the shock propagates through the specimen. To obtain zero-pressure densities comparable with those from synthesis experiments, complex extrapolations to atmospheric pressure must be made. It follows that densities from shock-wave experiments are less reliable than those from the equilibrium synthesis of high-pressure phases.

A simple example of reconstructive transformations in silicates is that of pure SiO_2. A high-pressure silica polymorph known as **stishovite** (after S. M. Stishov) was first synthesised at about 100 kbar pressure and was then recognised from meteorite impact craters. Stishovite has a zero-pressure density of $4280 \, kg \, m^{-3}$ and is characterised by octahedral coordination of silicon and oxygen (six oxygens squashed around each small silicon atom), whereas normal low-pressure silica, or quartz, has a density of $2650 \, kg \, m^{-3}$ and tetrahedral coordination (four oxygens around each silicon). Although there is unlikely to be free silica in the mantle assemblages described earlier, some may be produced by reactions in the Transition Zone and stishovite may be significant. The tendency to increase the coordination number for oxygen surrounding cations due to increasing pressure is a common theme in the Transition Zone. In spinel, however, the coordinations of magnesium and silicon are octahedral and tetrahedral, respectively, just as in olivine, but there are different bonds in the more closely packed, dense spinel structure.

The conditions of pressure under which natural olivines (the $Fe_2SiO_4 - Mg_2SiO_4$ series) transform to the spinel structure are shown in Figure 7.12. (Note that this new spinel is not $MgAl_2O_4$ or $MgCr_2O_4$ which were present higher in the upper mantle; the actual *volume* occupied by a single unit of Mg_2SiO_4 is rather less than for $MgAl_2O_4$ and $MgCr_2O_4$.) The olivines in upper mantle materials are magnesium-rich (80–100% Mg_2SiO_4), and consequently the olivine-to-spinel phase change (Fig. 7.12) will occur at 110–130 kbar pressure, or about 400 km depth. The 400 km discontinuity (Fig. 7.11) has, therefore, become firmly linked with this transformation.

Some of the other reactions that occur in the Transition Zone are quite complicated because of the many different structural states involved. Table 7.3 (p. 127) lists the important low-pressure compounds whose structural states are mimicked by mantle silicates at high pressure. Chemical formulae given on the left of Table 7.3 refer to the mantle minerals, whose structural states at increasing pressure appear on the right. (The interested reader can follow up these atomic structures by referring to the reading list.)

Although Mg_2SiO_4 is more compact in the spinel structure than as olivine, further contraction is possible if there is an increase in cation coordination number. In olivine and spinel, magnesium is in octahedral (six-fold) coordination and silicon is in tetrahedral coordination. There are two possibilities:

(a) A change from spinel to the structure that is characteristic of strontium plumbate (Sr_2PbO_4) at the Earth's surface.

125

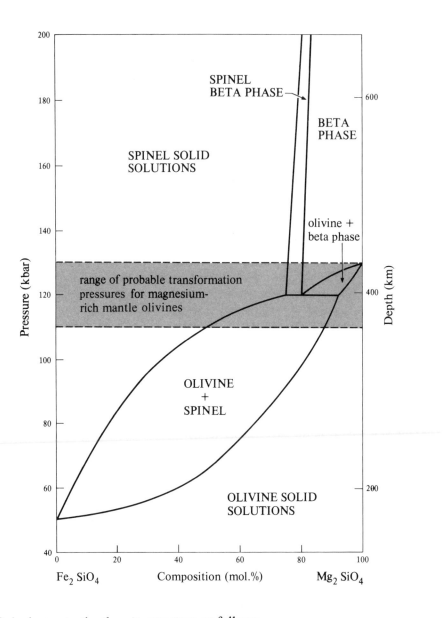

Figure 7.12 Phase diagram for Fe$_2$SiO$_4$–Mg$_2$SiO$_4$ at 1000°C. The beta phase is a spinel-like mineral possessing similar density but more complex crystallography than true spinel. (After Ringwood & Major 1970. Used with the permission of Elsevier Scientific Publishing Co.)

(b) A change to the *ilmenite* structure as follows:

$$\underset{\text{spinel structure}}{Mg_2SiO_4} \longrightarrow \underset{\text{ilmenite structure}}{MgSiO_3} + \underset{\text{periclase}}{MgO} \qquad (7.4)$$

Of these (a), which leads to a 10% increase in zero-pressure density, is favoured marginally over (b), with an 8% density increase, for the 650 km discontinuity. A third possibility is that complete separation may occur into simple oxides; for example, stishovite and periclase (Table 7.3).

There is even less certainty about the phase changes that may characterise the 1050 km discontinuity. Most workers accept that at least some upper mantle olivine must have reached the dense, simple oxide state but there is also the possibility that MgSiO$_3$ may pass from the ilmenite to postulated *perovskite* or *calcium ferrite* structures with about 7% and 10% increases in zero-pressure density, respectively.

Table 7.3 Chemical formulae and structural identity of possible polymorphs in the Transition Zone.

Chemical formula	Possible structural forms with increasing pressure*
SiO_2	quartz $\rightarrow$ stishovite
MgO	periclase
FeO	wüstite
Al_2O_3	corundum $\rightarrow$ ilmenite ($FeTiO_3$)
Mg_2SiO_4	olivine $\rightarrow$ spinel ($MgAl_2O_4$) $\rightarrow$ strontium plumbate (Sr_2PbO_4)
$MgSiO_3$	pyroxene $\rightarrow$ ilmenite ($FeTiO_3$) $\rightarrow$ perovskite ($CaTiO_3$)
$Mg_3Al_2(Si_2O_6)_3$	aluminous pyroxene $\rightarrow$ garnet ($Mg_3Al_2Si_3O_{12}$)
$NaAlSi_3O_8$	feldspar $\rightarrow$ jadeite ($NaAlSi_2O_6$) $\rightarrow$ calcium ferrite ($CaFe_2O_4$)

* Chemical formulae are quoted for low-pressure minerals of equivalent structure where they differ from the formulae given at the left of this table.

Note also that, in most silicate minerals, magnesium can be replaced by any proportion of iron in solid solution and also, in some cases, by calcium. In the main, these elements follow the same scheme through the mineral structures quoted.

So far, only the transformations of olivine, the most abundant upper mantle phase, have been considered. Also significant in garnet lherzolite above the Transition Zone are *garnet and pyroxene with subordinate feldspar*. The pattern of change for these minerals is analogous to those in olivine and is summarised in Tables 7.3 and 7.4. The following boxed section of text discusses some of the changes and it can be omitted at the reader's discretion.

For **pyroxene** ($MgSiO_3$), the logical high-pressure change is to the ilmenite structure with a 10–15% density increase, depending on composition. But there may be an intermediate stage where the garnet structure is used, even in cases where no aluminium is present:

$$2MgSi_2O_6 \rightarrow Mg_3(Mg,Si)Si_3O_{12} \quad (7.5)$$

and the (Mg,Si) group occupies the aluminium sites of ordinary garnet. If the pyroxene does contain aluminium, as do many lherzolite pyroxenes, then normal garnet will form directly just as it did from normal spinel higher up in the mantle:

$$\text{aluminous pyroxene} \rightarrow \text{garnet} + \text{stishovite} \quad (7.6)$$

To balance this equation, some SiO_2 is produced (stishovite) and will be used up in olivine–pyroxene conversions. Again, about a 10% density increase applies to pyroxene–garnet transformations, and experimental evidence suggests that this accompanies the change from olivine to spinel (structure) at the 400 km discontinuity.

Garnet is a stable phase across this discontinuity and is more abundant, at the expense of pyroxene, between 400 and 650 km depth. But at pressures over 200 kbar it transforms to the ilmenite structure with an 8% density increase. Even aluminium, which normally forms the oxide *corundum*, may well enter the ilmenite structure (Table 7.3). However, calcium, a relatively minor element in the mantle, which enters pyroxene ($CaSiO_3$) and garnet but not olivine, is more likely to produce a perovskite than an ilmenite structure. But both $MgSiO_3$ and $FeSiO_3$ are thought to pass through an ilmenite stage below 650 km and only enter the perovskite structure at higher pressures, possibly at 1050 km depth.

The net result for pyroxene and garnet is the same as for olivine transformations, except that Al_2O_3 is involved. This has prompted the suggestion that $MgSiO_3$ may recombine with Al_2O_3 at the bottom of the Transition Zone producing $MgAl_2O_4$, which now has the calcium ferrite structure, and stishovite. A mixture of perovskite, calcium ferrite and stishovite structures at 1050 km depth is thought to be about 7% more dense than a mixture of simple oxides; both are calculated on the basis of the same lherzolitic element abundances. As to which resulting mixture, mixed oxides or complex structures, is the more likely

127

match to available density solutions, much depends on the composition of the lower mantle (Section 7.5).

To complete this discussion, one other minor component of lherzolite deserves mention because of its sodium content (Table 7.2): this is plagioclase, which contains $NaAlSi_3O_8$. Beneath the **plagioclase** lherzolite layer this becomes part of the chemically complex pyroxene molecule, as $NaAlSi_2O_6$. High-pressure pyroxenes with this chemistry are known from crustal rocks where they are called **jadeite**. Sodium is incompatible in the garnet structure and so, as pyroxene is converted into garnet, jadeite may survive as a Transition Zone mineral until eventually it breaks down to produce $NaAlSiO_4$ in the calcium ferrite structure, together with SiO_2.

7.4.2 Mineralogy–depth scheme in the Transition Zone

Table 7.4 provides a summary of the possible changes in mineralogy with depth in the Transition Zone. Zero-pressure densities are quoted because experimentally synthesised materials are measured at atmospheric pressure. At the top of the Transition Zone, zero-pressure and actual densities are quite similar but, at deeper levels, this becomes less true.

Table 7.4 Summary of the major phase changes in the Transition Zone.

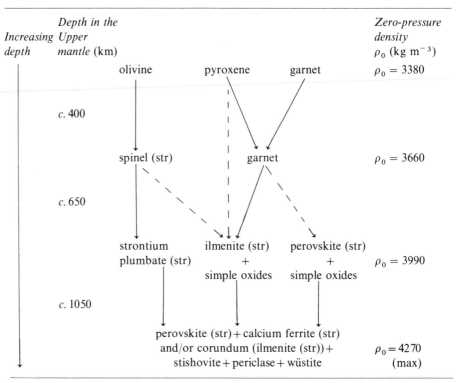

(str) = structure. (See the items marked * in Table 7.3.)
Dashed lines indicate alternative routes.

Ringwood (1969, 1975) suggested that the minerals in Table 7.4 are present at different depths in the following abundances. From 150 to about 400 km depth the minerals are:

128

Olivine	57%
Orthopyroxene	17%
Aluminous clinopyroxene	12%
Garnet	14%

At about 400 km depth, olivine transforms to the spinel structure and pyroxene to garnet, both over a range of several tens of kilometres due to the variable compositions (Fig. 7.12). From about 450 to nearly 650 km, density increases are due only to self-compression and the uniform material comprises:

Spinel structure	57%
Garnet	39%
Jadeite pyroxene	4%

These proportions assume a constant chemical composition for the different parts of the Transition Zone. All the olivine above the 400 km discontinuity has been converted to spinel and all the pyroxene, except for sodium-bearing jadeite (Table 7.3), has been converted to garnet. The total increase of density across this discontinuity is about 8%.

The next discontinuity occurs at about 650 km depth where spinel reorders to the strontium plumbate or, less likely, to the ilmenite structure. The magnesium, iron and aluminium silicates of garnet transform to the ilmenite structure, but its calcium silicate component adopts the perovskite structure. Jadeitic pyroxene transforms to the calcium ferrite structure. The result is another uniform region between 700 and 1050 km and, using the shorthand '(str)' for 'structure', this will comprise:

Strontium plumbate (str)	55%
Ilmenite (str)	36%
Perovskite (str)	6·5%
Calcium ferrite (str)	2·5%

The SiO_2 released in the jadeite–calcium ferrite (str) transformation goes to convert some strontium plumbate (str) to ilmenite (str). The zero-pressure density is now $3990 \, kg \, m^{-3}$ which is identical to the mean density of an oxide mixture in lherzolite proportions, with SiO_2 as stishovite and Al_2O_3 as ilmenite. This represents about a 9% density increase across the 650 km discontinuity.

Possible transformations across the small 1050 km discontinuity (Fig 7.11) are the conversions of ilmenite and strontium plumbate structured material to perovskite ($MgSiO_3$) and calcium ferrite ($MgAl_2O_4$) structures. The possible existence of these structures in the mantle is based entirely on inferences from shock-wave and analogue studies. A 7% density increase across this discontinuity is possible if these phase changes run to completion, giving a final zero-pressure density of $4270 \, kg \, m^{-3}$ at the top of the lower mantle.

At the probable pressure–temperature conditions of 1050 km depth this density becomes $4820 \, kg \, m^{-3}$, rather greater than the accepted value of $4600 \, kg \, m^{-3}$ (Fig. 3.14). Mixed oxides would give a low result, around $4500 \, kg \, m^{-3}$. This leaves some uncertainty about the state and composition of the materials at this depth and we return to this subject below. Notice, however, that in developing the model in Table 7.4 we have made the important assumption that there is no significant change in chemical composition across the Transition Zone.

129

7.5   The lower mantle

The lower mantle extends from 1050 km down to 2885 km and, according to Figure 3.14, density increases linearly from $4600 \, kg \, m^{-3}$ to $5500 \, kg \, m^{-3}$. The only indication of non-linearity occurs in a transition region where the mantle abuts the core. Here, S-wave velocity decreases with depth (Sections 2.3 and 8.6.2), and this may reflect the presence of high temperature gradients.

Birch (1961) made the first significant attempt to deduce the composition of lower mantle materials, characterised through the mean atomic weight ($\bar{m}$), which he tried to relate empirically to P-wave velocity and density (see Section 3.6 and Note 3). This is the molecular weight divided by the number of atoms, so $\bar{m}_{SiO_2} = 20 \cdot 0$, $\bar{m}_{MgO} = 20 \cdot 2$ and $\bar{m}_{FeO} = 35 \cdot 9$. For an assemblage, $\bar{m}$ is the mean of values for its constituents and equal amounts of these three oxides would give $\bar{m} = 25 \cdot 4$. For most common rocks, $\bar{m}$ is about 21, *increasing with iron content* by about one unit for every 6% FeO. Birch was unable to distinguish between several plausible compositions, but he did conclude that the lower mantle is a relatively homogeneous region in which density increases are due only to self-compression.

This approach has been developed with the aid of shock-wave experiments; for example, Ringwood (1969) (also Liebermann & Ringwood 1973) quoted a value of $\bar{m} = 21 \cdot 34$ for pyrolite and, *assuming* that the mantle is homogeneous, there would also be a value of $\bar{m} = 21 \cdot 34$ in the lower mantle. He determined high-pressure densities for three assemblages (see Fig. 7.13):

(a) dunite with an Fe/Mg ratio giving $\bar{m} = 21 \cdot 34$;
(b) an Al_2O_3–Fe_2O_3 mixture with $\bar{m} = 21 \cdot 34$; and
(c) a mixture of SiO_2, MgO and FeO in pyrolite proportions.

Mixtures (a) and (c) ignore the minor chemical constituents of the upper mantle, but mixture (b) makes some allowance. Figure 7.13 shows that the densities of feasible oxide assemblages with $\bar{m} = 21 \cdot 34$ are *too low* for lower mantle conditions. There are two ways in which the density might be increased.

(a) It may be increased by *introducing complex high-density mineral structures* such as perovskite and calcium ferrite at 1050 km (Tables 7.3 and 7.4). The experimental evidence for such structures was not produced by the transient shock-wave experiments illustrated in Figure 7.13, but density increases in the lower mantle of up to 7% can be postulated if complex structures replace simple dense oxides. The alternative,
(b) is the possibility that *the mantle is not chemically uniform.* Various suggestions have been made, most notably that the lower mantle contains a more iron-rich mixture of higher mean atomic weight than the upper mantle.

A proponent of chemical change is Press (1968), who used his density solutions (Ch. 3), together with an equation of state (cf. Birch's law), to predict $\bar{m}$ in the upper and lower mantle. Successful models for the upper mantle gave $22 \cdot 0 < \bar{m} < 22 \cdot 8$ (rather higher than pyrolite) and for the lower mantle, $23 \cdot 8 < \bar{m} < 24 \cdot 6$. Using these figures, Press predicted Fe/Mg ratios of 0·25 for the upper mantle (cf. 0·2 in many nodules,

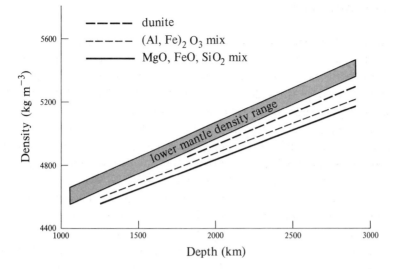

Figure 7.13 High-pressure densities for three assemblages in which $\bar{m} = 21.34$ compared to the lower mantle density distribution of Figure 3.14. (Data for mantle assemblages from shock-wave experiments of Ringwood (1969).)

Legend in figure:
- - - - dunite
- - - - $(Al, Fe)_2 O_3$ mix
——— MgO, FeO, SiO_2 mix

lower mantle density range

etc.; Section 7.3) and 0·6 for the lower mantle. This increase in mean atomic weight would raise the oxide mixture densities in Figure 7.13 into the observed range of lower mantle densities without the need for complex structures. Also in support of a compositional change, an increase of iron content is one means of accommodating all the iron of a chondritic Earth: chondritic silicates have Fe/Mg ratios of 0·45. However, this is a poor constraint because chondritic silicates are relatively oxidised and the Earth has an iron-rich core.

The most significant test for the compositional change hypothesis has come with the current debate about whole mantle convection (Section 8.6). If the lower mantle is more dense than the upper mantle, this would act to inhibit large-scale circulation; conversely, a compositional boundary could hardly exist if convection occurs. Ringwood (1975) has also argued that the ratio of bulk modulus to density (Eq. (3.14) Section 3.4) of a high Fe/Mg ratio lower mantle is inconsistent with seismic observations. During the late 1970s, these arguments have thrown the consensus back towards possibility (a) – the introduction of complex structures in the lower mantle – to satisfy density constraints while maintaining a homogeneous, lherzolite mantle. However, a total change to the most dense structures postulated may not be necessary. This is because only a 3–5% density increase over that of mixed oxides is needed (Fig. 7.13), whereas a maximum of 7% may be available. To summarise, there is a state of considerable ignorance about the composition and crystal structure of the lower mantle, but the best estimate we can make is that lherzolite chemistry is a mantle-wide feature. To account for the density of the lower mantle, it is likely that some structural forms of very high density are required in preference to a mixture of simple oxides.

Summary

The major sources of evidence used to determine the composition and state of the mantle are (i) seismic velocity and consequent density distributions, (ii) the results of high-pressure–temperature experiments on possible mantle materials,

131

(iii) ophiolite, kimberlite and other nodular materials that sample the upper mantle and (iv) meteorite data.

The following conclusions have been made:

1 Apart from the top few tens of kilometres, the whole of the mantle probably has a chemical composition like that of lherzolite, a kind of peridotite that, at the Earth's surface, comprises about 60% olivine, 30% pyroxene and 10% feldspar.

2 This material undergoes major partial melting (a maximum of about 25%) in the top 30 km of the mantle under spreading ridges where thermal gradients are high (about $100°C\,km^{-1}$), leaving harzburgite and dunite (olivine-rich) residues which are found at the bottom of upthrust ophiolite sequences.

3 At the lower thermal gradients elsewhere (about $10–15°C\,km^{-1}$), more restricted partial melting (5–10%) of lherzolite produces alkali basalt. Here, the intersection of the thermal gradient curve with the *wet* melting curve for peridotite below 50 km is probably responsible for the seismic low-velocity zone.

4 Relatively undepleted upper mantle lherzolite assemblages contain spinel ($MgAl_2O_4$) between 50 and 100 km depth, and garnet ($Mg_3Al_2Si_3O_{12}$) at greater depths. These are sampled by spinel lherzolite nodules which occur occasionally in basalt and by nodules in kimberlite pipes, respectively.

5 At 400 km depth, phase changes from olivine to spinel structure and from pyroxene to garnet structure occur. At 650 km transitions of garnet to ilmenite and perovskite structures, and of spinel into the strontium plumbate structure take place. These phase changes cause considerable increases in density because of structural changes in the atomic frameworks of minerals without any change in chemical composition. But the degree of confidence in these conclusions decreases with increasing depth.

6 A further small density increase occurs at 1050 km depth, where lower mantle mixtures are produced. The alternatives are mixed simple oxides in high-density structures, or complexes of compounds such as $MgAl_2O_4$ and $MgSiO_3$ in calcium ferrite and perovskite structures, respectively. Mixed oxides are unable to produce high enough densities unless the mantle is inhomogeneous on a grand scale with a much higher Fe/Mg ratio in its lower than in its upper part. This scale of heterogeneity is not widely accepted and so the complex structure model is favoured. These phases probably persist across the lower mantle, which is a region of relatively smooth seismic velocity and density increase with depth.

Further reading

General books:
Harris (1971); Clark (1971): broad basic review of mantle composition.
Wyllie (1971): summary of basalt magma genesis and other aspects of the mantle.
Advanced journals:
Cann (1974): mantle melting and production of ocean crust.
Harris & Middlemost (1970): review of kimberlites.
Clark & Ringwood (1964): summary of the pyrolite model

Birch (1961); Ringwood (1969); Press (1968); Davies (1977c): State and composition models for the Transition Zone and lower mantle.

Advanced books:

Ringwood (1975): discussion of most aspects of Chapter 7.

Carmichael *et al.* (1974): summary of basalt magma genesis.

Coleman (1977); Gass & Smewing (1980): reviews of ophiolites.

133

8 The dynamic mantle

8.1 Introduction

It has long been understood that mountain building involves great vertical movements, but the underlying causes were obscure until the discovery of plate tectonics, in which the vertical movements occur as a by-product of much larger horizontal motions. However, this only pushes understanding back a stage: What are plates? And what moves them? This leads our enquiry into how internal energy sources can generate forces and how the Earth responds. It will be shown that some sort of convection is involved.

A related question is: how far does the Earth depart from spherical symmetry? To an excellent approximation, it consists of concentric layers, but the division of the surface into oceans and continents is one example of departure from symmetry. We wish to know how deep the ocean–continent difference extends, and whether other **lateral inhomogeneities** exist. This topic is related to convection, because convection necessarily involves lateral variations in density.

8.2 Vertical movements and isostasy

The first evidence for large vertical movements of the crust came from measurements involving gravity. As recounted in Section 1.3, attempts to 'weigh' the Earth (*c*.1740) by comparing the gravitational pulls of its mass and that of a mountain showed that the mountain had less mass than expected. Later, this was found to be a common feature of mountains, including the Himalayas, but an explanation had to wait until 1855, when Sir George Airy and Archdeacon Pratt independently formulated the principle of **isostasy**. This is that mountains are high only because they are made of lighter material 'floating' on a denser and yielding substrate. Their explanations differed as to whether the highest mountains are highest because they are composed of the lightest rocks (Pratt, Fig. 8.1a), or because they are thickest (Airy, Fig. 8.1b), but both agreed that mountains, as a whole – and, indeed, the whole surface layer of the Earth – float on a denser material. The concept of isostasy evidently requires a rigid layer over a plastic one: the upper layer must have finite rigidity to preserve the topography of the Earth, for otherwise the mountains would level themselves like oil poured on water, while the lower layer must yield to permit the floating. The rigid and plastic layers are termed, respectively, the **lithosphere** and **asthenosphere.**

Of course, the finite strength of the lithosphere means that it can support some changes of load without making isostatic adjustments, but on a global scale this strength can be neglected.

Pratt and Airy had few ways of testing their theories, but now it is known that, though both types of isostatic compensation operate, mountains are high primarily because they are thick and so have deep 'roots', whereas continents, as a whole, are higher than ocean floors because they are composed of less dense material. That is to say, there is *a fundamental difference in structure and composition between oceans and continents*, a theme that will be returned to in this and the next chapter.

The existence of the plastic asthenosphere may seem to be incompatible with the transmission of S-waves which can propagate only in an elastic rigid material. The resolution of this paradox depends upon the *time scales*: the asthenosphere is rigid to stresses lasting only a short time but yields to long-continued ones, as described in the following section.

gravity uniform along here because mass in a column below this line is the same everywhere

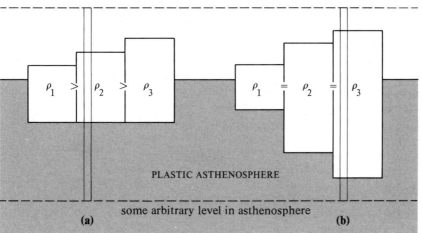

PLASTIC ASTHENOSPHERE

some arbitrary level in asthenosphere

(a)   (b)

Figure 8.1 Isostatic compensation: (a) according to Pratt's hypothesis, (b) according to Airy's hypothesis. If compensation is perfect, the masses in similar columns through the lithosphere down to the same level in the asthenosphere are identical.

8.3   The viscosity of the mantle

A material, such as that of the mantle, which can behave sometimes as a rigid solid, sometimes as a plastic one, obviously has complex mechanical behaviour, but it is permissible to discuss its viscosity as if it were a liquid, provided we realise that it applies only to deformation caused by steady stresses acting for long periods.

True viscosity is a measure of the 'thickness' or resistance to flow of a liquid. To deform a liquid it must be sheared, and we can think of a series of infinitely thin layers sliding over each other (Fig. 8.2). As the layers slide past each other, molecules jump back and forth between layers, tending to equalise the velocities of the layers and – more important for our discussion – the bonds that exist between the molecules, and prevent them dispersing as a gas, have constantly to be broken and remade. These processes produce the viscosity. In the simplest liquids, such as water, the rate at which the layers slide past each other is simply proportional to the shear force ('Newtonian viscosity'), but many liquids behave in a more complex fashion. Egg white and thixotropic (non-drip) paint are two familiar examples.

Now let us turn to the shearing of crystalline solids. In a crystal, atoms are held in positions that are fixed relative to neighbouring atoms by the forces between them (Fig. 8.3a). Consequently, if a moderate shear force is applied, the lattice as a whole distorts (Fig. 8.3b). If the force is removed, the lattice returns exactly to its original shape, and it is said to be behaving **elastically**. But if the stress is increased sufficiently, the bonds will break. In the laboratory, this may well cause the sample

135

to fracture into two, but, deep in the Earth, the surrounding pressure is so great that the crystal is always held together. In this case, one possible result is as in Figure 8.3c. Clearly, the new crystal will not revert to its original shape if the stress is removed, so now there is **inelastic** or **plastic** deformation.

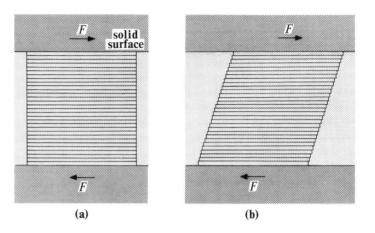

Figure 8.2 Shear in a liquid. The liquid can be thought of as consisting of many layers that slide over each other when sheared like a pack of cards.

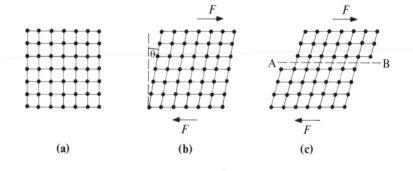

Figure 8.3 Shear in a crystal: (a) the undistorted lattice, (b) an elastic deformation due to a moderate stress. In (c) it is shown how a permanent inelastic deformation can result from a large stress.

The wholesale sliding of one part of a crystal past the other occurs only rarely. This is because all natural crystals have imperfections and these permit smaller adjustments to take place locally within a crystal, sometimes involving only one atom at a time. Many adjustments made successively throughout a crystal can add up to a significant permanent deformation. This is termed **creep** and is well known to engineers. In materials at room temperature, creep slows as the easier adjustments are 'used up' ('work hardening'). But the increased atomic mobility at temperatures approaching the melting points permits creep to continue indefinitely, and this applies to all the mantle below a few tens of kilometres (Fig. 7.3). An analogous example of creep in a crystalline solid is the movement of glaciers down their valleys. Under these conditions a solid behaves as a very stiff liquid and a viscosity, η, can be determined from the relation for Newtonian viscosity:

$$\text{shear stress} = \eta \times \text{rate of strain} \qquad (8.1)$$

Consequently, the mantle can support elastically an extra stress, if it is applied only for a time short compared to that required for atoms to make jumps and other permanent adjustments. This explains, at a crystalline level, how a material can behave viscously under a steadily applied force and, at the same time, elastically under a short-lived one. *Thus, to the passage of seismic waves which last only seconds, the mantle is an elastic solid, but, to crustal loading and other forces persisting for many thousands of years, it is a viscous liquid.*

The detailed atomic adjustments that permit creep to occur can take many forms, depending upon the temperature, the size of the stress, and the nature of the material. In some mechanisms of creep, the creep rate is highly dependent upon the magnitude of the stress, perhaps to the fourth power, whereas in others it may be simply proportional. At present, there is no agreement on which of several possible mechanisms is dominant in the mantle, and, because of this, we cannot be sure that our present limited knowledge of viscosity is applicable to all conditions that can exist in the Earth.

The viscosity of the mantle can be deduced as the result of some experiments Nature has carried out for us. The building-up and melting of several kilometres thickness of polar ice-caps have provided a sufficient long-term change of stress to deform the crust and mantle. Because of the high viscosity of the mantle, the isostatic adjustment following melting has been slow and is still taking place.

Before these adjustments can be used to deduce a viscosity, we need a model for the viscous behaviour of the crust plus mantle. The usual one used is a three-layer model: rigid lithosphere over an asthenosphere of relatively low viscosity, in turn over a lower mantle of relatively high viscosity. The argument for a lower-viscosity layer immediately below the lithosphere comes from Figure 7.3, which shows that, in this region, the temperature is closest to the melting point and, indeed, probably produces partial melting (but see Section 8.5). Of course, a three-layer model is a simplification, since the layers will grade into each other, but it is sufficient for our purposes.

To gain an intuitive understanding of how the model will respond to loading, consider the example of a thin elastic sheet of wood floating on thin oil, in turn resting on thick oil, each layer being denser than the one above (Fig. 8.4). The behaviour of such a system will depend upon the size of the load. A small load will be supported by the wooden sheet alone; a somewhat larger load will cause the wood to bend until the load is supported partly by the strength of the sheet and partly by the buoyancy of the wood depressed into the oil; for very extensive loads, we can neglect the strength of the sheet. Since the oil has to flow away before the sheet can deform, adjustment takes some time, and the higher the viscosity of the oil the longer this will be. The effect of the lowest layer, of thick oil, will be discussed shortly.

If these insights are applied to the Earth, we should expect small loads to be supported, without deformation, by the strength of the lithosphere. In fact, a substantial proportion of the weight of mountains, up to some tens of kilometres across, can be carried by Precambrian basement rocks, though not all rocks are as strong as this. However, there is no doubt that great ice-sheets, which can be hundreds of kilometres across and several kilometres thick, are far too heavy to be supported in this way and so depress the continental lithosphere as isostatic compensation occurs. There is direct evidence of this in Greenland, where the rock surface below the ice is saucer-shaped, with the centre below sea level. In the past,

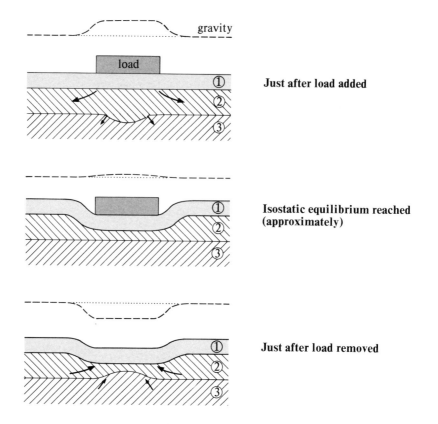

Figure 8.4 Schematic response of three-layer model to loading and unloading. For details see text.

Scandinavia was covered by a similar thick layer which started to melt 10 000 years ago, and surveying shows that the land is still rising, by as much as one centimetre a year, as it returns to isostatic equilibrium (Fig. 8.5). Gravity measurements relative to a predicted value for an Earth in equilibrium play an important part in these investigations, for the size of a **gravity anomaly** gives a measure of the present departure from isostasy, and supplements measurements of rates of uplift.

To apply our model to the real Earth, estimates of the thickness and extent, and hence of the weight, of past ice-sheets are deduced from detailed geological mapping. Mapping can also be used to deduce the amounts of subsequent uplift and when this occurred. The thicknesses and viscosities or strengths of these layers of the model are then adjusted to give the best agreement with these observations, for a range of ice-sheets of different sizes, with the results shown in Table 8.1. According to Cathles (1975), the middle layer is only 75 km thick, but what is more important is that the layer beneath is only about 25 times more viscous, which is not sufficient to prevent the flows that provide isostatic compensation extending into it. For this reason, the term asthenosphere strictly should apply to all of the region below the lithosphere; however, following common usage, its use will be confined to the least viscous layer just below the lithosphere.

The structure below the oceans can be deduced in a similar way, because the increase and decrease in the mass of continental ice-sheets produced changes in sea level of hundreds of metres and hence a significant change in the weight of water. The structure is similar to that below continents, with oceanic lithosphere tending to a thickness, as it ages (see Section 8.4), of about 100 km, only a little less than of

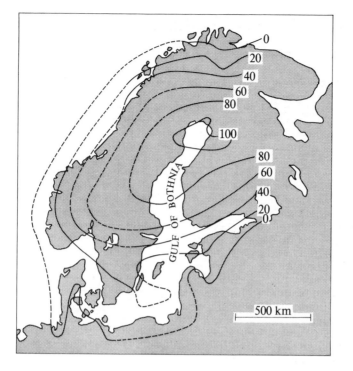

Figure 8.5 Uplift of Scandinavia. The contours show rates of uplift, in centimetres per century, deduced from surveys carried out since 1892. (After Gutenberg 1959, by permission of Academic Press.)

Table 8.1 Viscosity and rigidity of the mantle (which satisfy the three-layer model).*

Depth (km)	Division	Viscosity ($m^2 s^{-1}$)	Rigidity (N m)
0 – 100	lithosphere	—	5×10^{24}
100 – 175	asthenosphere	c. 4×10^{19}	—
175 – 2885	rest of mantle	c. 1×10^{21}	—

*Data from Cathles (1975).

continental lithosphere. However, this is not to deny that some lateral variations of viscosity may occur, for the method is not precise enough to detect them.

The surprisingly low viscosity of the lower mantle is a very important piece of evidence in favour of convection occurring in the whole of the mantle (Section 8.6.1), so the evidence upon which it is based will be discussed briefly. It might be thought that a thin layer of relatively low viscosity just below the lithosphere would be sufficient to accommodate the flows of mantle material caused by depression or uplift of the lithosphere, regardless of the viscosity at greater depths. This is not so, because of the huge extent of some of the ice-sheets; for example, the Wisconsin ice-sheet that covered North America was 2000 km across and several kilometres thick. If all the material displaced by the sinking of this vast load had to squeeze through a layer only 75 km thick, the result would be an increase in pressure in the layer, and this would lead to *uplift* of the unloaded area immediately surrounding the load (A and A′ Fig. 8.6). The converse would happen when the ice melted, which was the more recent event. Geological evidence is firmly against such bulges, which indicates that plastic deformation extends deep into the mantle.

139

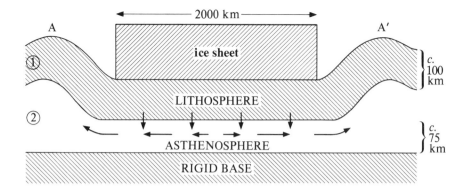

Figure 8.6 Schematic response of two-layer model with rigid base. The ice thickness is exaggerated. Geological evidence is firmly against this model.

8.4 What is a plate?

The accepted cause of mountain chains, oceanic ridges and trenches and some other topographic features is the relative movements of plates. These are conceived as more-or-less rigid pieces that cover the surface of the Earth, with tectonic processes largely confined to their edges. As this is not a book primarily about plate tectonics, a knowledge of these processes is assumed. Here the problem is what determines the thickness of a plate.

By definition, *a plate is all that mass which moves as a unit along the surface of the Earth over a yielding mantle.* It might therefore seem obvious that a plate is just a large continuous piece of lithosphere, but though this is true for oceanic plate, it is less simple for continental plate.

An oceanic plate forms at an oceanic ridge and thickens as it moves away. A close analogy is the freezing of ice upon a still pond: surface water solidifies as it gives up its latent heat of fusion to the colder atmosphere, but, as the ice thickens, the temperature *difference* between its upper and lower surfaces remains constant, so that the vertical temperature *gradient*, and hence heat flow through it, must decrease. Since the rate at which the water freezes onto the base depends upon the rate at which latent heat can be removed by conduction, the rate of **underplating** of ice must slow too. The same argument holds for oceanic plate and, since age is proportional to distance from the ridge axis, thickness will increase with this distance (Fig. 8.7).

Figure 8.7 Schematic section of an oceanic spreading ridge. The plate is bounded by the 0 and 1200°C isotherms. The arrows denote material being added by underplating. Because the plate material is cooler than asthenospheric material at the same depth, it is denser, and so sinks to maintain isostasy: columns A and B contain the same mass. Compare with Figure 7.10 for a petrological model.

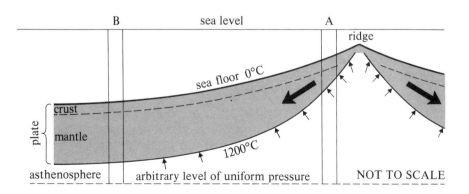

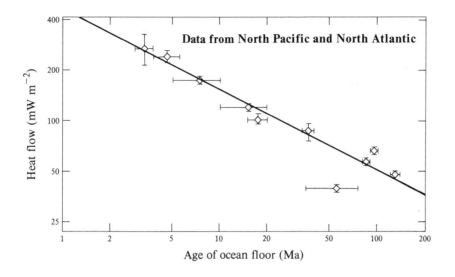

Figure 8.8   Heat flow versus age of ocean floor. The measured values lie close to the straight line, which shows that the heat flow is inversely proportional to the square root of the age of the ocean floor. Note that both scales are logarithmic. (Based on Sclater & Crowe 1979.)

It is not possible to confirm this model by measuring the plate thickness by direct seismic means, because the lower surface is somewhat arbitrarily defined – from laboratory experiments on the strength of the material – as roughly the 1200°C isotherm, and it is not a sharp transition from rigid to soft material. However, surface waves (Section 2.4) support a thickness increasing with age. The model can also be tested by its implications. First, heat flow should decrease with age, since the increasing thickness of plate increases its insulation. Simple theory shows that the heat flow should decrease as the square root of the plate age, and this is obeyed closely in practice (Fig. 8.8). Secondly, the ridge is high only because the material below it is hotter and hence less dense than elsewhere at a similar level. For instance, column A in Figure 8.7 consists mostly of hot asthenosphere, whereas column B is mostly cooler plate. Confirmation of this comes from gravity measurements which show that ridges are close to isostatic equilibrium (Fig. 8.9); this means that there is no large mass excess due to the ridge. Thirdly, it follows that the shape of a ridge depends primarily upon the coefficient of expansion of the material and upon the temperatures

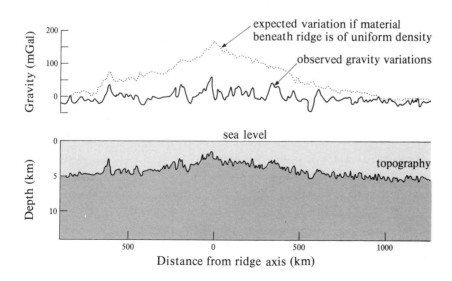

Figure 8.9   Topography and gravity of the Mid-Atlantic Ridge. *Lower*: observed topography; *upper*: observed and calculated variations of gravity, *g*. (Based on Talwani *et al.* 1965.)

141

below the surface. Plausible estimates of these quantities can be made and give good agreement with actual ridge sections, for an oceanic plate that thickens to about 100 km after 100 Ma (Sclater & Franchetau 1970; Sclater *et al.* 1980).

The behaviour of continental plates is not explained so simply. This is partly because they are less homogeneous than oceanic plates, being composed of rocks that vary in composition and thickness, and with temperatures that depend partly upon when last they were involved in orogenic or metamorphic processes. Moreover, it is known that differences between oceanic and continental areas extend to depths of several hundred kilometres. One source of evidence for this is seismological, using methods described in Section 8.9, and Jordan (1978, 1979) claims that differences exist down to at least 400 km, though other scientists contend that it may be only 200 km. Since the thickness of continental lithosphere, as deduced from the isostatic studies described in Section 8.3, is only about 100 km, this poses the problem of how a continental plate can translate as a rigid unit even though it includes the asthenosphere. The answer to this is not known: one possibility is that the asthenosphere can yield easily only a certain amount, which is sufficient to accommodate the relatively small vertical displacements required for isostatic compensation, but not the thousands of kilometres of lateral translation needed for plate movement. This is a topic where a better understanding of creep in the mantle is needed. However, the difference between oceanic and continental mantle is primarily compositional, rather than physical (Sections 7.3 and 9.3).

Because the lithosphere is not necessarily synonymous with plate, Jordan has proposed that the term **tectosphere** be used for the top part of the Earth that moves rigidly as a plate, reserving 'lithosphere' for the layer that behaves rigidly in isostatic movements. In the case of oceanic areas, the terms are equivalent, but this is not so for continental areas, the tectosphere being considerably the thicker.

8.5 The asthenosphere and the low-velocity zone

The asthenosphere is defined as the layer below the lithosphere which yields viscously during isostatic adjustments. The low-velocity zone is defined seismically (Section 2.3): it has a velocity reduction of 3–6% and is more pronounced for S- than for P-waves and, in addition, the attenuation of seismic waves passing through it is a few per cent higher than in the rest of the mantle. Though the asthenosphere and the low-velocity zone are formally distinct, they are often taken to be equivalent because they occur over roughly the same depths. However, this identification is an oversimplification.

The most popular explanation of the low-velocity zone is that it is a region of partial melt. Figure 7.3 shows how this arises where the mantle temperature exceeds the solidus when water is present. Estimates of the amount of melt needed to account for the low velocity layer vary from 0·1 to 6%. Laboratory experiments indicate that – contrary to intuition – these amounts of melt would not greatly affect the viscosity, which depends upon solid-state creep.

Thus, the mechanisms responsible for the existence of the asthenosphere and for the low-velocity zone are probably quite distinct. However, the asthenosphere and low-velocity zone have in common that they both depend upon processes that become important as the temperature approaches or reaches the liquidus, which is why they are found at roughly the same depth.

That the asthenosphere, in practice, is not the same as the low-velocity zone is shown by evidence that the latter is much less marked or even absent under continental shield areas, though isostatic adjustment shows that the asthenosphere exists there. Electrical conductivity measurements, which are believed to be sensitive to the presence of partial melt, also support the idea that partial melt is absent beneath shield areas (Drury 1978), probably because of the lower temperatures there (see Fig. 7.2 and Section 8.6.2).

8.6 The thermal régime of the mantle

8.6.1 *Convection*

The movement of plates requires some system of horizontal forces that can cause plates to collide and combine, or to break up, and the only possible mechanism seems to be convection.

The essence of convection is motion produced by buoyancy, with lighter material rising and denser sinking. This buoyancy can arise from the separation of material of different densities, as has already been described in connection with nickel–iron alloy partitioning in the core (Section 6.3.2). However, any separation in the mantle at present is no more than minor (Section 10.1), so that, if significant convection occurs, the density differences must be due to temperature differences: that is, it must be *thermal convection*.

Convection is usually illustrated by the homely example of soup heated in a pan, and though convection in the Earth must be far more complicated than this it can make clear that convection necessarily requires *lateral* density variations. Suppose that, before the soup is heated, it is confined in vertical columns by means of tubes, as shown at AB and CD in Figure 8.10. Next, the temperature distribution that exists in heated soup is established in some unspecified way: hottest up the centre and coolest at the top surface and sides. Column AB will expand and rise to A′. The pressure at A is now greater than that at C, even though they are at the same level, so that if a hole is drilled at A liquid will flow out sideways. This tends to lower the

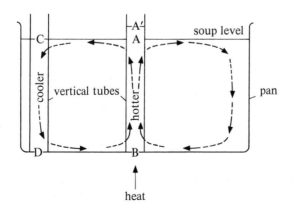

Figure 8.10 Density analysis of convecting soup. For details see text.

liquid in the tube so that the pressure at B, until now equal to that at D, will be less, and this pressure difference will tend to move liquid from D to B. Remove the tubes and liquid will circulate as the arrows show, due to the pressure difference arising from the lateral density differences.

Because of the possible complexity of convection, only the simpler cases have been studied closely. One of these is Rayleigh–Benard convection which occurs in a tank of liquid when the base is heated uniformly and the surface is cooled uniformly. Gentle heating only establishes a uniform vertical temperature gradient, and heat is transported upwards by conduction. Even though the lower layers are less dense, being hotter, than the upper layers, there is no *lateral* variation of density, and convection does not occur. Increased heating increases the buoyancy of the lower layers and there comes a point that, as soon as any slight lateral irregularity develops, lighter material will rise. Since irregularities develop spontaneously, convection occurs automatically. At first, it takes the form of regular convection cells roughly as wide as the depth of the liquid and fitting together like the cells of a beehive (Fig. 8.11). As in the soup pan, hot material ascends at the centre and cooler material descends around the edges of each cell. Increased heating, as well as increasing the vigour of the system, progressively concentrates flow to the axis and surfaces of each cell. With further increases in basal heating, the regularity of the cells begins to disappear: they begin to move about and change in size, and finally the pattern gives way to irregular flow with columns or blobs rising at random.

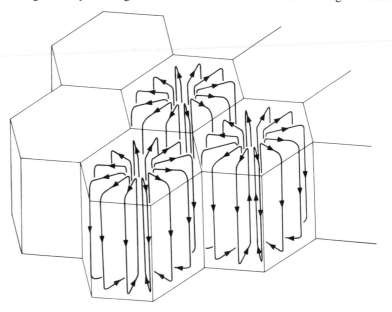

Figure 8.11 Rayleigh–Benard convection.

Lord Rayleigh showed that, in general, convective behaviour depends upon a dimensionless constant, now known as the Rayleigh number:

$$Ra = \frac{\text{(volume coefficient of thermal expansion)} \times \text{(vertical temperature difference)} \times g \times \text{density} \times \text{(depth of liquid)}^3}{\text{thermal diffusivity} \times \text{viscosity}}$$

or, in symbols

$$Ra = \frac{\alpha \, \Delta T \, g \, \rho \, d^3}{K \, \eta} \tag{8.2}$$

Increasing the values of the quantities in the numerator encourages convection: α because it is the thermal expansion which lowers the density and so creates buoyancy, g and ρ because the difference in the weight of hot and cold columns of equal height is proportional to the pull of gravity and the average density, d because the taller the columns the greater the difference in pressure between the tops of the columns (A and C in Fig. 8.10). The effect of increasing ΔT is obvious, though it is not the actual temperature difference between top and base, for we saw in Section 3.5 that, in a liquid compressed by the weight above, there is no buoyancy unless the temperature exceeds the adiabat, i.e. the temperature gradient due to compression. Since the mantle can be regarded as a compressible liquid so far as convection is concerned, ΔT in Equation (8.2) is the vertical temperature difference (between top and bottom of column) in *excess* of the adiabatic difference. Increasing the quantities in the denominator discourages convection: the thermal diffusivity, K, given by conductivity/(density × specific heat), is a measure of the heat in the hotter, buoyant parts of the liquid that is lost by conduction, and so an increase in K reduces the temperature gradients that are essential for convection, whilst an increase in the viscosity, η, obviously will tend to inhibit any motion in the liquid.

For a simple system, such as a homogeneous liquid heated as its base, the Rayleigh number for convection to begin is very roughly 2000, the actual value depending upon the shape of the system and other details. This is the condition for the onset of convection, but for convection to be vigorous and the major transporter of heat, with conduction making little contribution, the Rayleigh number must be much larger, say 10^5, whilst above 10^6 convection is likely to be irregular. For the whole mantle, approximate values of the quantities are estimated to be: $\alpha = 2 \times 10^{-5}\,°C^{-1}$, $g = 10\,ms^{-2}$ (g is nearly constant throughout the mantle, see Fig. 3.14), $\rho = 3500\,kg\,m^{-3}$, $d = 3 \times 10^6\,m$, $K = 10^{-6}\,m^2\,s^{-1}$ and $\eta = 10^{21}\,kg\,m^{-1}\,s^{-1}$ (Section 8.3). Thus $Ra \approx 2 \times 10^4 \Delta T$. ΔT is not well known, but even if it were only $1°C$, convection would occur; more probably, ΔT is several hundred degrees (see following subsection), so that convection is likely to be vigorous and even irregular. However, as this conclusion is based upon simpler systems than the mantle, it is only a strong probability and not a certainty.

The detailed form that convection takes will depend upon the complexities of the mantle, which include the existence of phase changes, the movements of plates and the distribution and size of the heat sources. At present, mathematical analysis is not able to handle all of these variables (which often are known only poorly), while laboratory models, though valuable, cannot incorporate such features as phase changes or spherical symmetry. Thus, only broad conclusions can be drawn at present.

Two particular aspects of convection are so important that they will be discussed separately, in Sections 8.8 and 8.9: these are plumes and the forces that drive plates. But first, the heat sources and temperature distribution in the mantle will be discussed.

8.6.2 *Heat sources and the temperature in the mantle*

If we knew how heat sources and temperature are distributed in the mantle, we should be considerably nearer knowing what form convection takes, but current knowledge of these topics is very poor. Models have to be constructed based upon the general knowledge of the Earth already described in this book, and tested against the few available observations.

One observation that all models have to satisfy is the total surface heat flow. The value of the heat flow (Note 9 relates how this is measured) varies over the surface of the Earth, as already described for oceans (e.g. Fig. 8.8); continental variation will be discussed in Section 9.3.2. In addition, measurements are distributed unevenly over the Earth, so global totals are somewhat uncertain. A value of roughly $3 \cdot 2 \times 10^{13}$ W was calculated by Lee and Uyeda (1965). However, it is claimed that this neglects a significant amount of heat removed by circulating water at and near oceanic ridges, and that the value should be increased to roughly 4×10^{13} W (Williams & Von Herzen 1974).

There are many possible ways in which heat can be produced in the mantle, but only three are potentially big enough to need consideration here. First, the mantle already contains sufficient stored heat – by virtue of heat produced in the past – to supply the present surface heat flow for several times the age of the Earth. To do this it would, of course, have to cool, so the question is whether it is actually cooling. The surprising conclusion is that it is probably cooling only slightly, because of the stabilising effect of convection. This arises because, once a system is heated sufficiently to be convecting, a further increase in heat input primarily results in more vigorous convection, rather than an increase in temperature gradient. Thus, little of the extra heat is stored, for nearly all is delivered to the cold surface. Conversely, a convecting system will not convect for long after heat input has ceased, and so the Earth's mantle is convecting not because of past, but largely

Table 8.2 (a) Selected heat flow data.

	Continental average (W m^{-2})	Oceanic average (W m^{-2})	Total (W)
through surface of Earth	53×10^{-3} *	62×10^{-3} *	3×10^{13} * 4×10^{13} †
through Moho	28×10^{-3} *	57×10^{-3} *	$2 \cdot 4 \times 10^{13}$ *
through core–mantle boundary	—	—	$0 \cdot 4$–$1 \cdot 6 \times 10^{13}$ ‡

*Pollack and Chapman (1977).
† Williams and Von Herzen (1974).
‡ Elsasser *et al.* (1979).

Table 8.2 (b) Estimated radioactive contents and heat productivities (concentrations are taken from Tables 9.1 and 9.2).

	Uranium, U (ppm)	Thorium, Th (ppm)	Potassium, K (%)	Total heat production (μW m^{-3})
average continental crust	1·6	5·8	1·7–3·0	1·0–1·1
average oceanic crust	0·9	2·7	0·4	0·5
undepleted mantle	0·015	0·08	0·1	0·02

Note: a heat productivity of about $0 \cdot 03$ μW m^{-3} through the whole of the mantle could supply the observed surface heat flow of the Earth.

because of present, heat input. This brings us to the second source, heat from the core. This is inferred because an energy source is needed to drive the geomagnetic dynamo. As described in Section 6.2.3, this may be due either to decay of ^{40}K or to growth of the inner core, both of which must deliver heat to the base of the mantle. Estimates are very uncertain but range from 0·4 to 1·6 × 10^{13} W (Table 8.2a), with a value somewhat less than 10^{13} W being favoured. This is a significant but not major fraction of the total surface heat flow, which leads to the third source, radioactive decay within the mantle.

The amount of radioactivity in the mantle is a matter of controversy at present. Theory suggests that since the major radioisotopes (isotopes of U, Th and K) are strongly lithophilic, they should be concentrated near the surface of the Earth. This seems to be the case, especially for uranium and thorium where the amounts in the crust approximately equal the total complement of a chondritic Earth (Section 5.4.1), though up to about 20% of the expected potassium is in the crust. It has been suggested that the remaining potassium may be in the sulphide zone of the core (Section 6.2.3), but this is by no means certain and so the Earth may not contain the chondritic amount. The fact that the current heat production of a chondritic Earth would closely equal the observed total surface heat flow has been taken as evidence for the Chondritic Earth Model, but this is an oversimplification because of the possibility of significant heat contributions from stored mantle heat and from the core, neither of which need derive from radioactive heating. Attempts to deduce the radioactive content of the mantle from mantle-derived samples are limited by the relatively shallow depths from which the samples derive, and by their widely varying contents of radioactive elements. However, Table 8.2b gives estimated concentrations of radioactive elements in mantle and crust. The mantle is highly depleted in the radioactive elements – by a factor of about 200 – compared to the continental crust, but because of its great bulk their total heat productivities are probably similar, and suffcient to make up the observed total surface heat flow. Thus, the surface heat flow is chiefly attributed to radioactive heat production in the continental crust and the mantle, with a contribution from the core, which may also be radiogenic. In addition, there may be a smaller amount due to mantle cooling.

Though continental crust is both richer in radioactive elements and thicker than oceanic crust, the average surface heat flows for oceans and continents are similar (Table 8.2a). Therefore, the heat flows up through the Moho in the two regions must be very different. Pollack and Chapman (1977) have used the estimated radioisotope contents of crustal rocks to subtract the crustal heat production, in effect stripping the crust from the mantle. They concluded that the heat flow through the oceanic Moho is about twice that through continental Moho (Table 8.2a), which is consistent with the idea that the continent–ocean difference extends to great depths (Sections 7.3 and 8.4).

As explained in Section 8.4, oceanic heat flow is determined by the heat that is conducted up through the thickness of plate from the convecting mantle beneath; heat production by radioactive decay within the plate is quite small. The lower heat flow through continental Moho depends upon two factors. Because the continental plate or tectosphere is much thicker (up to 400 km, or more), heat transfer by the relatively slow process of conduction is much reduced compared to the thinner oceanic tectosphere, i.e. it acts as a thicker layer of insulation. Secondly, the continental tectosphere beneath the Moho is considerably depleted in radioactive elements compared to similar depths below oceans. This has been brought about by their upward concentration into the continental crust, together with other lithophilic elements (Section 9.2.2). This depletion results in a more refractory material with low heat productivity, which explains how the mantle deep beneath continents can be rigid enough to translate as a plate, whereas the suboceanic mantle at comparable depths cannot.

Putting together the ideas that have been discussed in Sections 6.3.2 and 7.2.1 and in this chapter, tentative temperature profiles can be deduced, and are given in Figure 8.12. In the tectosphere, temperature gradients are determined largely by conduction, though modified by *in situ* heat production. In the crust, the continental temperature gradient is steeper than the oceanic one because of the much higher concentration of radioactive elements. But, beneath the Moho, the continental gradient is less than that in oceanic mantle because it is highly depleted in these elements and because continental tectosphere is thicker. Even so, the gradient is still considerably steeper than in the convecting part beneath, where the gradient cannot greatly exceed the adiabat. At the base of the mantle, a thin layer with high gradient is likely because of the probability of considerable heat input from the core. (The non-convecting layers at the top and base of the mantle are often referrred to as thermal boundary layers, and are commonly found in convecting systems.) Such a high gradient would account for the small increase in seismic velocities found there (Section 2.3). The temperature at the base of the mantle must match that of the top of the core, which is estimated to be about 3000°C, and certainly in the range 1800–3900°C (Section 6.3.2).

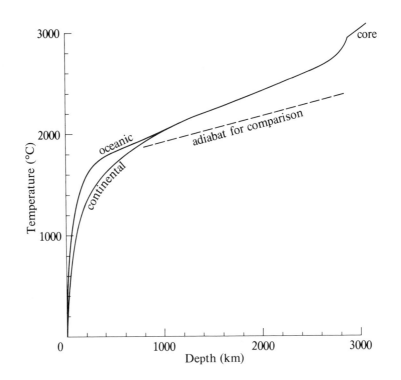

Figure 8.12 Possible temperature profiles in the mantle.

8.7 Hot spots and plumes

The majority of the world's volcanoes occur along plate boundaries, in association with spreading ridges and subduction zones. However, some volcanism is also found within plates in small isolated areas, and these have been called hot spots.

Examples of hot spots on land are Yellowstone in the U.S.A. and Tibesti in the Sahara; in the oceans there are many isolated volcanoes formed away from ridges. Some of these hot spots form linear traces of volcanism, though contemporary activity occurs only in a small area at one end. The best known of such traces is the Emperor–Hawaiian chain in the Pacific (Fig. 8.13). Two other chains in the Pacific are closely parallel to the Hawaiian chain and this led to the idea (Wilson 1963) that the sources of the hots spots may be fixed in the mantle and that the chains record the passage of the Pacific plate over them.

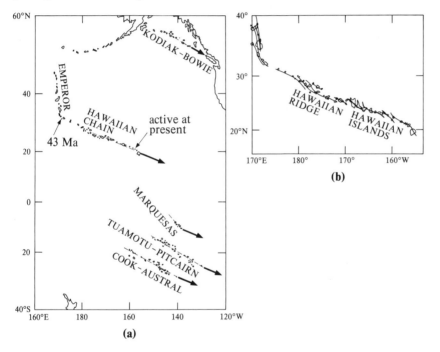

Figure 8.13 Volcanic island chains in the Pacific. (a) Present activity occurs only at the south-east end of each chain. (b) The smaller figure shows that the volcanoes have an *en échelon* arrangement. (Based on (a) Turcotte & Oxburgh 1978, by permission of the Royal Society; (b) a drawing by Turcotte & Oxburgh and reproduced by permission from *Nature* **244**, 333–9. ©1973 Macmillan Journals Ltd.)

Wilson has extended the concept to areas on a plate boundary which have higher activity than average. One such example is Iceland which is anomalous, for part of a spreading ridge, in being above sea level. As it has aseismic ridges that stretch east and west to the Faeroes and Greenland, it seems to have been an area of higher activity ever since the north-east Atlantic opened over 50 Ma ago.

It has been claimed that hot spots are the surface expression of plumes, which are thin columns of hot material, rising like smoke from a cigarette in still air. Experiments show that plume-like behaviour can occur with a wide range of viscosity contrast between plume and surroundings.

It is often suggested that plumes may originate in the thermal boundary layer at the base of the mantle produced by the inflow of heat from the core (Section 8.6.2). The layer is unstable because of its high temperature gradient, and blobs or columns of hot material will rise spontaneously. However, uneven delivery of heat from the

149

core, reflecting convection in the core, may tend to localise where material breaks away from the layer. But, though plumes might form in favourable circumstances, there are objections to their existence. One problem is how plumes can remain fixed for over 50 Ma in a mantle that is convecting. Though it is known that hot spots are not quite stationary, since they have relative motions of a few millimetres a year (about a tenth of the velocities of many plates), this still seems difficult to reconcile with whole mantle convection.

Attempts have been made to detect the hot, low-density column of a plume from the effect it should have upon seismic rays passing through it. Some years ago it was claimed that anomalous material exists below Hawaii, but now it seems the effects arise in other parts of the ray path (Green 1975) and, in general, there is little seismic support for plumes.

Chemical evidence in their support has also been proposed. The volcanism of hot spots predominantly comprises alkali basalts, whereas ridge basalts are mostly tholeiites, and, as most alkali basalts have deeper sources (Section 7.2.2), this has been attributed to the plumes bringing up material from greater depths. Associated with the different composition are differences in trace element composition of possible source materials and in strontium initial isotope ratios (Note 5), which argue for fundamentally different source regions that have been separate for a long period of time. What form this mantle heterogeneity takes is a matter of debate at present, but the topic will be discussed further in Section 9.3.3.

It has been suggested that plumes on plate boundaries may be driving the plates, much as the scum on soup is carried away from the hot upwelling column. Such forces have been included in some of the analyses of plate driving forces considered in the next section. The conclusions was that they make no contribution (Chapple & Tullis 1977) so, to the extent that such analyses are valid, this is evidence against the plumes, at least as driving forces.

Alternative explanations for some hot spots have been offered by Turcotte and Oxburgh (1973, 1978). They pointed out that, because an oceanic plate cools as it moves away from a spreading ridge, it will contract and so develop tensional stresses parallel to the spreading ridge. In general, this will tend to produce fractures at angles to the stress axis, and in this way Oxburgh and Turcotte have been able to account for the trend of the Hawaiian chain. Another cause of intraplate stress arises because the Earth is a flattened sphere, so that when a plate changes its latitude it has to change its curvature, and this will produce tensional stresses, as in a piece of orange peel when it is forced to lie flat. It has been calculated that both of these mechanisms could develop sufficient stress to fracture a plate, and such fractures could propagate with time, either as a plate cools or as it moves over the surface of the Earth, and the head of the crack might permit magma to ascend and form a hot spot. Tensional cracking seems a better explanation than a plume for the *en échelon* distribution of volcanism along the Hawaiian chain (Fig. 8.13), but cannot account for single small areas of volcanism, such as isolated volcanic islands.

In summary, the areas of volcanism away from plate margins, which differ in composition and strontium initial isotope ratio from ocean ridge basalts, are not accounted for by present plate tectonic theory. Plumes offer a superficially attractive explanation, but it is difficult to be quantitative about their properties, and they remain a rather vague concept for which there is little supporting evidence.

8.8 The forces that may move plates

Though the basic principles of plate tectonics were enunciated in the 1960s, there is as yet no agreement about what moves the plates. This is not for lack of ideas about possible driving forces, but rather because of the difficulty of testing their action.

The proposed forces acting on plates can be divided into edge and basal forces, some of which will propel, whilst others will retard motion (Fig. 8.14). Gravity sliding operates on the flanks of spreading ridges, and this is simply the sliding of the plate down the slope (Fig. 8.7). The moving-apart of plates may be assisted by wedging action as the magma rises between them (Lachenbruch 1973). The sum of the forces acting at or near ridges is termed **ridge-push** force.

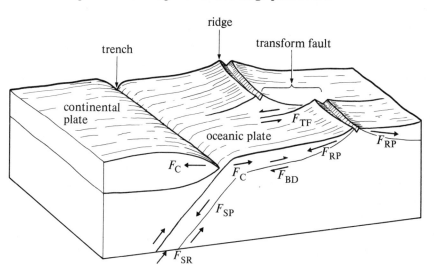

Figure 8.14 Forces acting on plates: F_{RP}, ridge-push; F_{TF}, transform fault friction; F_{BD}, basal drag or drive; F_C, collision force; F_{SP}, slab-pull; F_{SR}, total mantle resistance to descending slab. (Based on Forsyth & Uyeda 1975, by permission of the Royal Astronomical Society.)

Another driving force is the pull of a subducting plate, termed **slab-pull**. It arises because, although an oceanic plate is made of roughly the same material as the mantle beneath it – being formed largely by underplating – once it subducts it is considerably cooler (Fig. 8.15) and hence denser than its surroundings. As a result it sinks, pulling the attached horizontal part of the plate after it. Evidence in support of this comes from fault-plane solutions of earthquakes which occur in the descending slab (see Note 10), which show that the top of the slab is in tension. The existence of phase changes increases the negative buoyancy of the slab because, being cooler, the phase changes occur at a shallower depth than in normal mantle, accentuating the density contrast of slab and surrounding mantle. (Temperature, by tending to expand the lattice, tends to oppose a phase change which would lead to a higher density; see Fig. 7.9, for example.)

There is little doubt that drive by ridge-push and slab-pull forces occurs, but much more problematical is the sense of the force on the base of plates. A basal force will arise by viscous drag whenever there is relative motion between a plate and the mantle beneath. In a soup pan, the scum is moved by the motion of the convecting soup beneath, and this model has been applied to the plates. However, this need not be the case, because the ridge-push and slab-pull forces may be sufficient to move the plates, in which case the basal force will be retarding. The question, therefore, is

151

whether slab-pull and/or ridge-push provides the major plate driving force, or whether the major force is **basal drag** by the moving mantle. That is to say, do the plates drag the mantle, or does the mantle drag the plates?

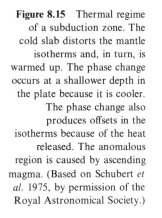

Figure 8.15 Thermal regime of a subduction zone. The cold slab distorts the mantle isotherms and, in turn, is warmed up. The phase change occurs at a shallower depth in the plate because it is cooler. The phase change also produces offsets in the isotherms because of the heat released. The anomalous region is caused by ascending magma. (Based on Schubert *et al.* 1975, by permission of the Royal Astronomical Society.)

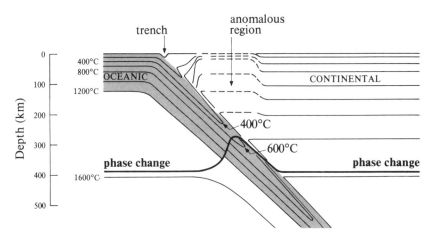

In addition to the driving forces, there must be retarding forces, for otherwise the plates would continue to accelerate. Retarding forces arise by friction where plates 'rub' past each other (F_C and F_{TF} of Fig. 8.14) and by viscous drag where a plate subducts into the mantle (F_{SR}), and perhaps basal drag as discussed above. Since the magnitude of the viscous retarding forces increases with the speed of the plates, the plates will move at a steady speed in which the combined drive forces are balanced by the sum of the retarding ones. This balance of forces makes it difficult to determine the magnitude of the driving forces acting on the plates.

Despite this, a number of attempts have been made to analyse the plate forces. One approach relies upon the fact that the various forces act in different proportions on the different plates. For instance, the Pacific plate has an area 30 times that of the Cocos plate (Fig. 8.16): if the dominant drive force is basal drag, with retarding forces acting only at the plate edges, then the Pacific plate should go faster since it has the greater ratio of base to edge. But, to apply this approach, it is necessary to know the 'absolute' plate velocity, i.e. with respect to the mantle beneath.

Various ways have been used to estimate 'absolute' plate velocities. Among them are the methods of assuming that hot spots are fixed in the mantle or of assuming, rather arbitrarily, that a certain plate or margin is at rest.

Most investigators have concluded, regardless of the way it has been deduced, that the absolute velocity of a plate depends chiefly upon its proportion of edge which is subducting: that is, *the major drive force is slab-pull* (Fig. 8.17). The correlation is so strong that it seems other forces are only minor, and it follows from this that the major retarding force must be the viscous resistance of the mantle to the subducting slab. If it were not so, there would also be a correlation of plate velocity with basal area or whatever determined the retarding force. Thus, it would seem that slabs descend at near their terminal velocities, just as a stone sinks through water at a speed where viscous drag balances its weight in water.

But the other drive forces cannot be negligible, because there are plates with little or no subducting edge, such as Africa and Antarctica, which nevertheless have

152

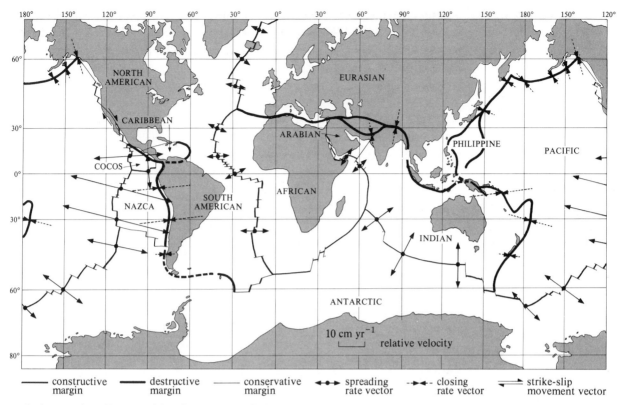

Figure 8.16 Plate map. The map shows 12 plates; other small or poorly defined plates also probably exist. The arrows show relative motions determined from sea-floor magnetic anomalies and earthquake recordings. (Based on Minster *et al.* 1974, by permission of the Royal Astronomical Society.)

relative motion. Presumably, the pattern of plate movement is determined largely by the fast plates, which are those with much subducting edge, while the remaining plates move around in the 'spaces between' under the action of the minor forces, such as ridge-push.

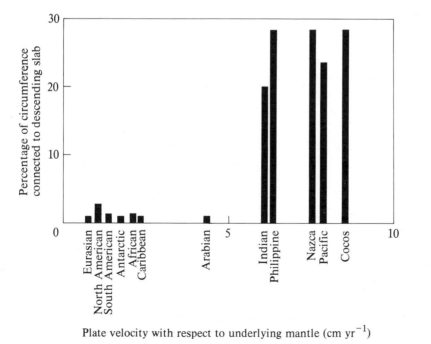

Figure 8.17 Plate velocities versus percentage of subducting edge. The strong correlation suggests subducting edges drive the plates. (From Forsyth & Uyeda 1975, by permission of the Royal Astronomical Society.)

However, these conclusions are not universally accepted, and different approaches sometimes give differing results. This disagreement may be resolved in the future with the help of more evidence, such as a detailed knowledge of the stresses within plates. These can be determined from fault-plane solutions of the relatively uncommon earthquakes that occur within plates, or in other ways.

The reader may be wondering what connection there is between plate driving forces and convection. The answer is twofold: first, that gravity-sliding and slab-pull *are* examples of convection and, secondly, that plate and adjacent mantle motions must have some effect on each other.

Thermal convection is merely the circulation induced by hot material rising and/or cold material sinking. In the case of a plate, the processes of hot material rising at the ridge, sliding down the ridge slope and then descending in a subduction zone all depend primarily upon buoyancy forces arising from temperature differences, and so are examples of thermal convection (Fig. 8.18a). Of course, there must be some form of broad, sluggish mantle flow to complete the convective flow, but the bulk of the drive appears to come from the descent of cold material in the plates. This contrasts with Rayleigh–Benard convection (Fig. 8.18b) where the concentrated flow is the hot material rising up the axes of the cells. These two examples demonstrate a more profound difference. In (b) the plates are carried along passively by the horizontal surface flow, whereas in (a) flow in the body of the liquid is initiated by the movement of the plates. Thus, if it is true, as suggested, that plates are driven chiefly by slab-pull, *mantle convection could be driven by the plates, and not vice versa*. This may seem strange to those familiar only with the soup pan analogy, but other examples of surface layers moved by forces largely independent of the liquid beneath are known. If a cauldron of molten metal is allowed to cool, a skin of solid metal forms on the surface, but being denser that the underlying liquid it will founder and slide into the molten liquid. On a larger scale, a lava lake in Hawaii has exhibited similar behaviour, for sheets of solid rock form on the surface and move about, exhibiting many of the features of plate tectonics; spreading ridges, transform faults and a form of subduction are all observed (Duffield 1972). In addition, calculation indicates that a plate moving at a few centimetres per year is able to induce flow that extends down through the whole of the mantle (Davies 1977a).

Figure 8.18 Two extreme convection models. In (a) the concentration of force and flow is in the plate, from ridge crest to the foot of the subducting slab. Mantle adjacent to the plate gets carried along by viscous drag. In (b) the concentration of force and flow is in the ascending column and the plates are carried along passively by the surface currents. Arrows give a rough idea of the distribution of velocities.

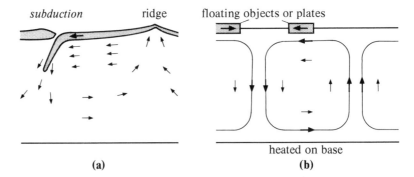

But mantle convection is not likely to follow either of these extreme examples and is probably something between. Although the probable heat sources deep in the mantle are sufficient to produce convection in the bulk of the mantle, the form the convection takes is determined partly by the boundary conditions, which include the positions of the plates. Probably, we should think of plates and the mantle as forming a single, though complex, system, with each affecting the other. An illustration is provided by a tank filled with glycerine in which convection is initiated by rising bubbles of cold carbon dioxide which, on reaching the surface, form a cold layer of gas resting on the glycerine (Fig. 8.19). A cold surface layer of glycerine forms which, solely because of its resulting higher viscosity and density, behaves rather like an oceanic plate, even subducting. It can be seen that the subducting plate affects flow in its neighbourhood, so that the flow in the tank is determined both by the rising bubbles of carbon dioxide and by the 'plate'.

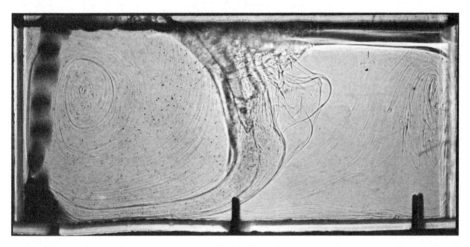

Figure 8.19 Glycerine tank. Bubbles of cold carbon dioxide rising, left, produce a circulation and also cool the surface, forming a cold 'plate' which descends at the centre. (After an experiment by J. S. Turner, 1973).

The realisation that plate movement need not be a passive reflection of mantle convection helps to explain the existence of the spreading ridges which almost surround Africa (Fig. 8.16). Since there is no subduction between the Atlantic and Indian Ocean ridges, they must be moving apart. If spreading ridges could form only above rising convection currents (Fig. 8.18b), this would imply that the separation of the Indian, African and South American plates is due to the changes in the positions and sizes of vast convection cells. But, if plates are moved chiefly by forces within them (Fig. 8.18a), ridges are merely the places where the lithosphere is always weakest because thinnest, and the flow of material from mantle into plate is just that needed to fill the gap between the separating plates and maintain isostasy, the release of pressure promoting partial melting (Section 7.2.2).

Clearly, there is much to be learnt about mantle convection, and the discussion here does no more than to illustrate some possibilities and approaches. But, at least, the reader should no longer believe that convection is necessarily like the behaviour of heated soup!

8.9 Lateral inhomogeneities

Lateral inhomogeneities may be detected by a variety of techniques, of which seismology is the most important. However, as the variations in seismic velocities

155

are small – less than 1% – sophisticated methods have to be adopted to detect them. One method is to compare the travel times of two rays that have paths roughly in common, except for the region of interest. Figures 8.20(a) and (b) illustrate two such cases. The method shown in part (a) is suitable for detecting deep-seated differences between continental and oceanic plate, and can be supplemented by comparing the dispersion of surface waves which have travelled over different parts of the Earth (Fig. 2.9). Both of these methods have shown the existence of inhomogeneities.

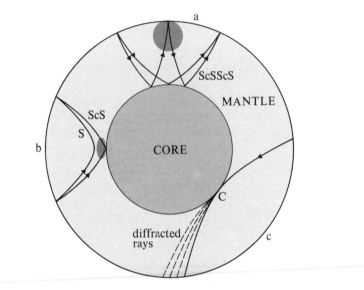

Figure 8.20 Three ways of detecting seismic inhomogeneities. In (a) and (b), if the shaded region is anomalous, the travel time of a ray passing through it will differ from the standard value. If the travel time of the companion ray – which is roughly in common except for the shaded region – is normal, the anomaly cannot lie anywhere along the path except for the shaded region. In (c) the amplitudes of diffracted rays are sensitive to inhomogeneities at C.

In addition, rays that travel through subduction zones suggest that slabs may descend below 700 km – the maximum depth at which they can be detected by earthquakes which occur within them – to at least 800 km, and possibly as much as 1400 km (Jordan 1975a). This implies that they can remain cool and hence strong enough to retain their distinct identity, though no longer brittle enough to produce earthquakes.

There also seem to be inhomogeneities at the base of the mantle. Evidence comes from comparison of, say, S and ScS rays (Fig. 8.20b), and also from rays diffracted at the core surface, whose amplitudes are very sensitive to inhomogeneities at their source. (Diffraction is the bending of waves around objects, on a scale comparable with the wavelength.) These inhomogeneities may be related to the probable high temperature gradient in the region (Section 8.6.2), and the possible source regions of plumes.

A different way to detect inhomogeneities is from *large-scale gravity anomalies*. Because of isostatic compensation, significant anomalies extending thousands of kilometres laterally can only exist if dynamic forces are present to oppose continually the compensation. One example, met earlier, is the build-up or melting of ice-sheets, but another possibility is the effect of the forces produced by convection. In fact, large negative anomalies, indicating mass deficit, are found at oceanic trenches, indicating that material is being removed there, and this is attributed to the sinking of the subducting slabs. However, the anomalies now being discussed are much larger than this in lateral extent, and are usually detected from the perturbations of satellite orbits (Section 3.3.2). Although anomalies are found, their interpretation in terms of convection is tentative at present, and the interested reader will find a brief discussion in the boxed section that follows.

Gough (1977) found that the geoid (the sea level surface, which is determined by the Earth's gravitational field) departs from that expected for a rotating ball of liquid, or reference spheroid, by some tens of metres. Somewhat surprisingly, the high and low areas form two interlocking zones, like the pattern on a tennis ball. Gough noted that the greatest *depressions* were over regions of fast spreading, such as the East Pacific Rise and the ocean south of India, where there is a 'hole' 100 m deep. Conversely, there is an *elevation* over the North Atlantic which is opening at a rate only a third of that of the East Pacific Rise, perhaps because the plates either side of the Mid-Atlantic Ridge lack subducting edges. There are also geoid elevations over the subduction zones of the East and West Pacific.

Kaula (1972) considered global gravity anomalies down to 1200 km across, and on this scale found quite close correlations with surface features, e.g. a negative anomaly over northern Canada, as would be expected for an area that has not fully recovered isostatically from the melting of its ice load. An interesting feature is that the negative anomaly over Antarctica is three times larger than would be expected for incomplete adjustment to a likely thinning of its ice-cap. Kaula suggested that since Antarctica is almost surrounded by spreading ridges, which therefore must be retreating from it, they may be creating a slight 'vacuum' or mass deficit as material is sucked in to add to the growing Antarctica plate.

Additional evidence for the existence of inhomogeneities comes from other geophysical methods but adds little, as yet, to our knowledge. Thus, though there is little doubt that lateral inhomogeneities exist in the mantle, at various depths and on various scales, it is not yet possible to interpret them in any consistent way, nor to identify them with the compositional inhomogeneities for which there is also evidence (Section 9.3.3). But as the study of inhomogeneities is still in its infancy, the next few years should see solutions to some of these problems.

Summary

Most of the topics of this chapter are still young, with many loose ends remaining. The major conclusions are as follows.

1   The existence of isostatic compensation allows the effective viscosity of the mantle to be determined: a rigid lithosphere about 100 km thick rests on an asthenosphere of roughly the same thickness, with the rest of the mantle only about 25 times more viscous than the asthenosphere. The thickness of continental lithosphere is probably little greater than that of old oceanic lithosphere (i.e. about 100 Ma old).

   The solid-state creep mechanisms responsible for this viscosity are not yet agreed upon.

2   The viscosity and other parameters of the mantle are such that vigorous convection probably occurs through all the mantle except the boundary layers at top and base; it may be so vigorous that it is irregular.

3   The distribution of heat sources in the mantle is poorly understood. The major source of heat is probably due to radioactivity which, on the whole, is concentrated towards the surface. However, its concentration in the top several

hundred kilometres of the mantle beneath continents is less than that beneath oceans, helping to account for a Moho heat flow half that beneath oceans. Heat flow from the core into the mantle is likely to be significant, but unevenly distributed, reflecting convection in the core.

4 The temperature profile of the mantle is far from linear: there is a steep rise in the rigid plate where heat transport is by conduction, a roughly adiabatic rise in the convecting mantle which extends down to a non-convecting boundary layer, heated by the core. The temperature profile below continents is lower than that beneath oceans, to a depth of at least 400 km (see Fig. 8.12).

5 Oceanic plates form at ridges and thicken by underplating to about 100 km. Continental plates can be much thicker, possibly up to 400 km, or even more, and include the asthenosphere, unlike oceanic plates which are bounded below by the asthenosphere. The outer part of the Earth that moves *laterally* in plate movements is termed the **tectosphere**.

6 The deep differences between oceans and continents must involve at least rigidity, temperature, heat productivity, density and composition in an intimate way. For continental tectosphere to be thicker than oceanic tectosphere requires a lower temperature to provide the necessary rigidity; in turn, this needs a lower heat productivity. But a lower temperature, by itself, would produce a greater density and hence positive gravity anomalies over the continents; since these are not observed, there must be compensating compositional differences as well. This different composition helps to increase the rigidity of the mantle beneath continents. The compositional change involves the upward removal of radioactive elements into the continental crust and hence the lower heat productivity.

How this could have been brought about will be discussed in Chapter 9.

7 The asthenosphere and low-velocity zone are distinct, the former's properties probably being due to solid-state creep, while the latter's most likely arise from small amounts of partial melt. The low-velocity zone does not exist (or is greatly reduced) beneath continental shields, whereas the asthenosphere does exist there.

8 Hot spots are not accounted for by plate tectonics, and plumes have been proposed instead. These narrow, concentrated flows of uprising material are thought to originate at the base of the mantle, but the evidence for their existence is not compelling.

9 Plate driving forces are still not understood. Edge forces undoubtedly exist and some are important, but the role of basal forces – drag or drive? – is not known. At present, the negative buoyancy of subducting edges seems the most likely major drive force. Also obscure is the relation between plate movements and convection. However, the relationship is almost certainly an intimate one, with oceanic plates being seen as an integral part of the convection cycle; passive convection 'à la soup pan' is most unlikely.

10 Mantle lateral inhomogeneities exist, but how the various types (including compositional) are related to each other or to convection is not yet clear.

Further reading

As many of the topics discussed in this chapter lack a consensus view among scientists few suitable review papers exist.

General journals:

McKenzie and Richter (1976): mantle convection.

Burke & Wilson (1976): mantle plumes.

Jordan (1979): evidence for continental plate extending to 400 km depth and methods of detecting lateral inhomogeneities.

General books:

Runcorn (1967): *Dictionary of geophysics*: isostasy.

Kittel (1976): creep.

Uyeda (1978): general account of plate tectonics.

Advanced journals:

Stocker & Ashby (1973): mechanisms of creep.

McKenzie & Weiss (1975); Davies (1977b); Elsasser *et al.* (1979): convection.

Sclater *et al.* (1980): review of heat-flow and crust–mantle thermal models.

Chapple & Tullis (1977); Minster & Jordan (1978); Davies (1977c): analyses of the forces acting on plates.

Morgan (1971, 1972): mantle plumes.

Jordan (1975a, b, 1978): thickness of continental lithosphere and detection of lateral inhomogeneities.

Advanced books:

Cathles (1975): review of mantle viscosity (especially Ch. 1 and pp. 267 *et seq.*)

Le Pichon *et al.* (1973): physical aspects of plate tectonics.

9 The continental crust

9.1 The nature of the Earth's crust

Although the Earth's crust is the most accessible and best studied part of our planet, it is by far the most complex in both its physical and its chemical nature. It contains a wide range of rock types from the relative chemical purity of many sedimentary rocks to complex chemical mixtures such as igneous basalts and granites. The crust is a solid, rigid layer that comprises the uppermost part of the plates and, from seismic data (Section 2.3), it is likely to be of lower density than the upper mantle peridotites beneath the Moho discontinuity. The layered structure and thickness of basaltic ocean crust (which is only 5–10 km thick) was described in Section 7.3. By contrast, the average thickness of the continental crust is about 35 km, though there are immense anomalous thicknesses under the larger mountain ranges – up to 70 km beneath the Andes and up to 90 km beneath the Himalayas, for example. Areas where the continental crust is of less-than-average thickness are also localised as in the Kenya Rift Valley.

This chapter is about the present state of the continental crust and the processes operating within and beneath it. The main questions it aims to answer are:

(a) What are the major surface and in-depth features?
(b) What rock types are involved and how did they originate?
(c) To what extent can these features and rock types be accounted for by modern crust-forming processes?

Today, the major manifestation of deep crustal activity in the continents occurs adjacent to regions of active plate collision and involves the melting and metamorphism of subducted materials (Sections 9.3 and 9.4). At destructive margins, the continental crust accretes new material by this igneous activity, but through a *two-stage* process because melting at spreading ridges has already fractionated basaltic oceanic crust from the mantle (Section 7.3). But we start our study of the crust with the distribution, nature and age of the rocks at the surface and, using seismic information, at depth.

9.2 Crustal structure and composition

9.2.1 *Major surface features*

The continental land masses and the floors of the ocean basins contrast strongly in both topography and age. The major topographic features (Fig. 9.1) range between the highest mountain (Everest, 8848 m) and the deepest ocean trench (the Marianas, 10 912 m). But the distribution is bimodal, with the continents (average height 0·87 km) and ocean basins (average depth 3·7 km) being quite distinct. This difference in level is because the continental lithosphere contains a much greater

160

thickness of low-density crustal material than does the oceanic lithosphere and so, because of isostatic compensation in the asthenosphere, the continents stand proud (Section 8.2).

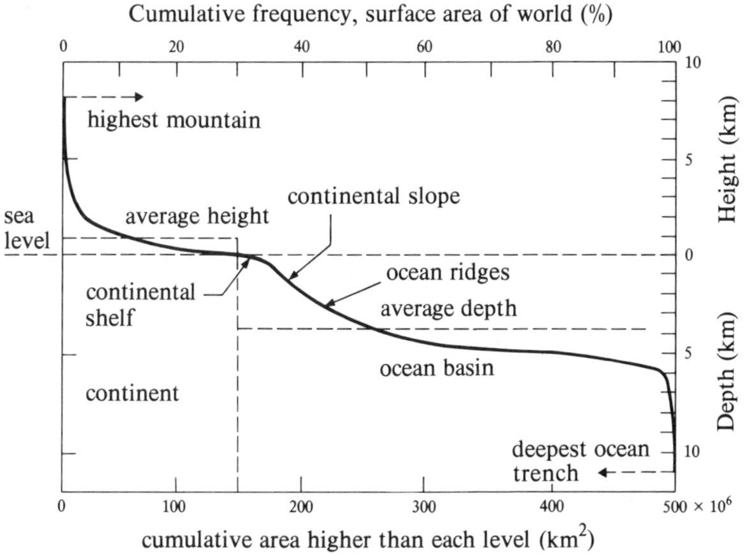

Figure 9.1 The distribution of levels in the Earth's solid surface on a cumulative basis. The total area of the Earth's surface is about 510×10^6 km², the average height of continental areas above sea level is 0·87 km and the average depth of ocean basins below sea level is 3·7 km. (Source: Wyllie 1971. Used by permission of John Wiley & Sons Inc.)

The oceans cover some 70% of the Earth's surface and they occupy large flat-bottomed basins that are traversed by 2 km high ridge systems (Figs 8.9, 8.16 and 9.2). Around their margins the ocean basins have trenches adjacent to destructive plate margins – as around most of the Pacific – or broad continental shelves, up to 200 km wide, where subduction is not taking place – as around most of the Atlantic. For simplicity, we shall refer to these as the **active** and **passive** edges of the continents. Because ocean crust spreads away from the ridge zone, it follows that the oldest ocean-floor rocks are found in the basin margins, but these rocks are no older than 200 Ma, and most were formed in the last 100 Ma.

In complete contrast, the continental crust comprises rocks formed throughout most of the 4600 Ma history of the Earth (Ch. 10) and, at the surface, these form three main components (Fig. 9.2):

(a) exposed continental **shields** that consist of Precambrian (> 570 Ma) crystalline igneous and high-grade metamorphic rocks;
(b) continental **platforms** where there is a gently folded cover of younger metasedimentary rocks, usually overlying more Precambrian basement; and
(c) young, mainly Tertiary (< 70 Ma) **fold mountains** that may contain older, deformed, metasedimentary rocks, and that almost always contain young igneous rocks, both volcanic and intrusive.

The continental shield and platform areas directly abut passive ocean/continent edges, such as those of the Atlantic (Fig. 9.2). Here, the *surface* transition from ocean to continent is gradual, with characteristic continental shelf zones which were produced by marine erosion during the recent (Pleistocene) glacial period of lower sea levels. The active margins of the Pacific are marked on the continental side by

161

volcanically active mountains, such as the Andes, or by island arcs, such as the Aleutians and Philippines (Figs 8.16 and 9.2). Another prominent young mountain belt occurs along a line extending east from the Mediterranean from the Alps to the Himalayas. Here, igneous activity is less prominent, but the degree of folding and deformation is extreme – slices of metasedimentary rocks are thrust over each other leading to considerable crustal shortening. The Alpine–Himalayan chain lies within the continental interior and is a zone of continent–continent collision that contrasts dramatically with the marginal mountains adjacent to the Eastern Pacific which mark the line of ocean–continent collision. In the latter case, relatively dense ocean lithosphere has been subducted beneath the lighter continental rocks and so there is much less evidence of deformation at the surface.

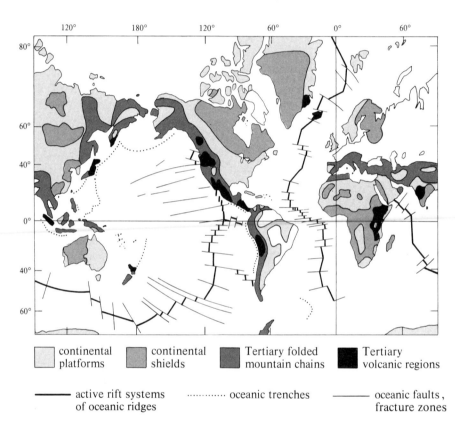

Figure 9.2 Major surface features of the Earth's continental crust in terms of their geological structure. See also Figure 8.16 for plate definitions. (Source: Wyllie 1971. Used by permission of John Wiley & Sons Inc.)

- continental platforms
- continental shields
- Tertiary folded mountain chains
- Tertiary volcanic regions

——— active rift systems of oceanic ridges   ············ oceanic trenches   ——— oceanic faults, fracture zones

9.2.2   *Vertical structure and composition*

The pattern of crustal thickness on a global scale is quite regular (Fig. 9.3). In continental areas, surface crustal divisions can be extended into the third dimension, for the depth of the Moho mirrors the surface topography in an exaggerated way: thus, the thickest crust occurs under the continental mountain ranges. Chapters 7 and 8 noted that the ocean–continent differences may well extend several hundred kilometres deep into the mantle. Above the Moho, the ocean crust has a prominent seismic layering (Fig. 7.4) related to petrologic variations among the basalts, gabbros and peridotites involved. Continental crust has a more complex layered structure, but the thickness and junctions between the layers are often far less well defined than in ocean crust.

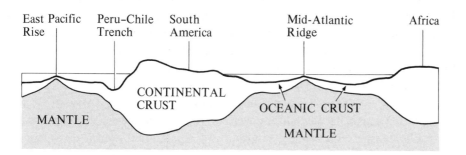

Figure 9.4 gives the seismic structure of the continents in terms of P-wave velocities for various contrasted regions (see also Fig. 2.9). Apart from the Moho, a more diffuse discontinuity occurs within the crust at depths ranging from 10 to 30 km, known as the Conrad discontinuity, after its discoverer. It can be traced under most continental regions and it divides the upper crust and lower crust which have P-wave velocities in the ranges 4–6.5 km s^{-1} and 6.5–7.8 km s^{-1}, respectively. The upper crust is usually the most variable layer with an uppermost few kilometres of material ranging from relatively unmetamorphosed volcanic or sedimentary rocks (profile (b) in Fig. 9.4, for example) to medium-grade metasediments (such as the quartzites and greenschists of profile (c)). Beneath this near-surface layer, the upper crust is typically characterised by a narrow range of P-wave velocities in the range 5.9–6.5 km s^{-1} and this is often regarded as 'granitic' in composition. However, acid plutonic rocks of the granite family vary considerably from intermediate diorites (about 55–60% SiO_2) to the rather less common granites *sensu*

Figure 9.4 Seismic structure of continental crust and upper mantle (V_p) in the following locations: (a) Wisconsin, a stable shield area; (b) Basin and Range province, a continental rift environment, where young volcanic rocks cover a thinned shield area; (c) northern Scotland a 400 Ma old continent–continent mountain belt; (d) southern California, a 100 Ma old ocean–continent mountain belt; (e) the central Andes, a modern ocean–continent convergence zone; (f) the central Alps, a modern continent–continent convergence zone. Light shading indicates sediment, metasediment and low-density crystalline rock velocities (4–6.5 km s^{-1}), intermediate shading indicates velocities (6.5–7.8 km s^{-1}) appropriate to high-grade crystalline metamorphic rocks, and dark shading indicates mantle peridotite velocities (> 7.9 km s^{-1}). (Data from P. J. Smith 1973, Garland 1971, Faber & Bamford 1979.)

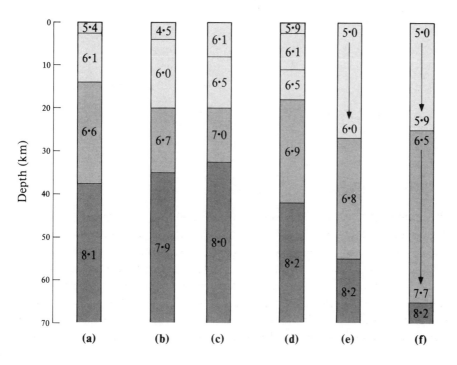

stricto (> 70% SiO_2). The latter typically have densities of 2650 kg m^{-3}, P-wave velocities of about 6·1 km s^{-1}, and are associated with negative gravity anomalies, indicating that the average density of the upper crust, like the P-wave velocity, is greater than that of granite (see P. J. Smith 1973). It is more likely that the *average* composition of the upper crust is granodiorite or quartz diorite with a density of 2700–2750 kg m^{-3} and an average P-wave velocity of 6·25 km s^{-1} (see also Section 9.2.3).

The composition of the lower continental crust is less well known because of the relatively few places where likely rock types are exposed. Traditionally, it was regarded as basaltic, but this idea became less popular with the realisation that some of a basaltic lower crust is under sufficiently high pressure to transform into dense *eclogite* (Section 7.2.1; see also Brown & Fyfe 1972). Although eclogitic rocks with the chemistry of basalt have very high seismic velocities of about 8·4 km s^{-1}, rocks with less basic compositions but in the eclogite metamorphic facies (i.e. formed under the same conditions of pressure and temperature – Fig. 9.5), may be present. Dioritic eclogite would contain quartz and feldspars with lower seismic velocities (V_p 7·3 km s^{-1}) than the garnets and pyroxenes that predominate in basic rocks at high pressure. But much of the lower crust has a lower velocity still and is also shallower than the mininum depth of 35 km (10 kb–Fig. 9.5) for the eclogite facies, and so other candidates must be sought.

A more widely recognised rock type for the lower crust is **granulite** which forms

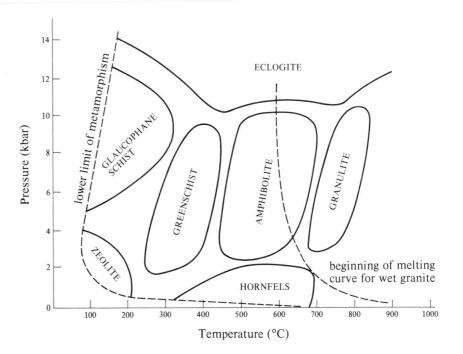

Figure 9.5 Simplified pressure–temperature scheme for metamorphic facies, also showing the beginning of melting curve for wet granite. (Adapted from F. J. Turner 1968.)

164

over a range of high-pressure and high-temperature conditions (Fig. 9.5). Granulitic rocks of intermediate-to-basic composition, containing mainly pyroxenes and calcium-rich feldspars have appropriate P-wave velocities of about $7\,km\,s^{-1}$. Yet a third possibility is that much of the lower crust lies in the field of **amphibolite** facies metamorphism. This would imply rather lower temperatures than the granulite facies (Fig. 9.5) and the presence of relatively hydrous rocks with amphibole and subordinate feldspar. As with granulitic assemblages, amphibolites would have appropriate velocities if their compositions lay within the intermediate–basic range (45–60% SiO_2).

Support for the occurrence of granulite and amphibolite facies rocks in the lower crust comes from their occurrence in the most deeply eroded crustal regions of Precambrian age. Granulite is most abundant and is favoured also for several other reasons. For example, heat flow studies indicate that the lower crust is depleted of heat-producing elements, as are many granulites (Heier 1978), and experimental work on granite melting indicates that, during the emplacement of acid magmas into the upper crust, the lower crust will be depleted of water, heat-producing elements and, probably, a silicate melt fraction, thus leaving only the refractory pyroxenes and feldspars of granulite (Sections 9.3 and 9.4). However, many originally granulitic rocks are known which have been rehydrated since they formed, perhaps due to remetamorphism at lower pressures and temperatures (retrograde metamorphism), so that they are now amphibolites.

Therefore, the most likely average composition and rock type of the upper crust is granodiorite, whereas that for the lower crust is granulite of intermediate-to-basic composition. However, the lower crust may also contain significant amounts of amphibolite, of similar composition, while occasional pockets of eclogite at deep levels cannot be ruled out.

9.2.3 *Gross composition of the crust*

We have seen that the Earth's crust can be divided into oceanic and continental types and that, on a global scale, both have layered structures. Before considering the processes of crustal growth and evolution, it is useful to summarise the chemical composition of the crust. Early attempts to determine the composition of the crust used weighted averages of all the varied rocks that have been collected at the Earth's surface. But this is invalid on two counts. First, the deeper layers were both poorly known and under-represented and, secondly, little attention was paid to the oceanic crust. Accurate estimates only became possible late in the 1960s when the granulitic nature of the lower continental crust was recognised and when the ocean crust was explored by drilling. Table 9.1 gives a breakdown of crustal composition by layer and by location (after Ronov & Yaroshevsky 1969). For example, the continental crust is subdivided into a small (about 7%) sedimentary layer, an upper crust of average granodiorite composition and a slightly more basic lower crust (see also Fig. 9.4). Recent estimates of the heat production in the lower crust indicate that potassium, in particular, may be a little too high in both the 'lower crust' and 'average continent' columns of Table 9.1. The overall intermediate and basic compositions of the continental and oceanic crust, respectively, are expressed clearly and they are separated by a 'subcontinental' category, comprising the

165

Table 9.1 The chemical composition of the Earth's crust and its component layers (wt%). (After Ronov & Yaroshevsky 1969.)

Layer	Continental crust			Oceanic crust			Totals		
	Sedi-mentary	'Upper'	'Lower'	Layer 1	Layer 2	Layer 3	Conti-nental	Sub-continental	Oceanic
mass of continents (%)	7·1	45·1	47·8	—	—	—	mass of all crust (%)		
mass of shelf/slope (%)	11·2	37·4	51·4	—	—	—	63·7	15·1	21·2
mass of oceanic crust (%)	—	—	—	3·1	15·9	81·0			
SiO_2	49·9	63·9	58·2	40·6	45·5	49·6	60·2	59·4	48·6
TiO_2	0·7	0·6	0·9	0·6	1·1	1·5	0·7	0·7	1·4
Al_2O_3	13·0	15·2	15·5	11·3	14·5	17·1	15·2	15·1	16·5
Fe_2O_3	3·0	2·0	2·8	4·6	3·2	2·0	2·5	2·5	2·3
FeO	2·8	2·9	4·8	1·0	4·1	6·8	3·8	3·8	6·2
MnO	0·1	0·1	0·2	0·3	0·2	0·2	0·1	0·2	0·2
MgO	3·1	2·2	3·9	3·0	5·3	7·2	3·1	3·2	6·8
CaO*	11·6	4·0	6·0	16·7	14·0	11·7	5·5	5·9	12·3
Na_2O	1·6	3·0	3·1	1·1	2·0	2·7	3·0	2·9	2·6
K_2O	2·0	3·3	2·6	2·0	1·0	0·2	2·8	2·8	0·4
P_2O_5	0·2	0·2	0·3	0·2	0·2	0·2	0·2	0·3	0·1
C	0·5	0·2	0·1	0·3	0·1	—	0·2	0·2	<0·5
CO_2*	8·3	0·8	0·5	13·3	6·1	—	1·2	1·5	1·4
S	0·1	<0·05	<0·05	—	—	<0·05	<0·05	0·05	<0·05
Cl	0·2	0·05	<0·05	—	—	<0·05	0·05	0·05	<0·05
H_2O	2·9	1·5	1·0	5·0	2·7	0·7	1·4	1·4	1·1

*The sedimentary continental layer and layer 1 of the ocean crust are high in CaO and CO_2 because of their calcite (limestone) content.

continental shelf and slope region, which is usually assigned to the chemically similar continental crust.

The analyses listed in Table 9.1 are expressed as oxides and, in terms of single elements, oxygen is the most abundant crustal component. This abundance of oxygen is associated with the upward concentration of lithophilic (oxygen-loving) elements compared with *chondritic* and *solar abundances* (Fig. 5.4). Together, the 10 most common elements listed in Table 9.1 comprise over 99% of the Earth's crust by weight and they are concentrated preferentially into the *mantle* partial melts that form beneath the ocean ridges and which create ocean crust (K and Al in Fig. 7.7, for example). However, a comparison of the garnet lherzolite (undepleted mantle) analysis of Table 7.2 with the oceanic and continental crust analyses of Table 9.1 shows that continental crust is by far the more enriched in strongly lithophilic elements. These geochemical differences are illustrated to good effect by many of the trace elements, which are more sensitive to melting and fractionation processes than major elements. In Table 9.2, a continuous progression of enrichment of the highly lithophilic elements, rubidium, thorium and uranium, can be seen from mantle, to oceanic crust, to continental crust, and vice

Table 9.2 The concentrations of some critical trace elements in the Earth's crust and mantle (all in ppm).

	Continental crust	Oceanic crust	Undepleted mantle*
Rb	70	30	0·6
Sr	400	465	34
Th	5·8	2·7	0·08
U	1·6	0·9	0·015
Ni	82	130	1500
Co	28	48	100

* Garnet lherzolite – see Section 7.3 and Table 7.2; data from Turekian (1972), O'Nions *et al.* (1978), Fyfe (1979) with modifications, see also Table 8.2.

versa for the chalcophilic/siderophilic trace elements, nickel and cobalt. (Strontium is also lithophilic, but, whereas it is incompatible in the structures of most mantle silicates (olivine, etc.) and enters preferentially into mantle melts, it happily substitutes for calcium in the feldspars of oceanic crust where it reaches a maximum concentration.)

All this evidence supports the idea that the continental crust has been formed from the mantle, either directly or indirectly, through a long history of melting events and the remainder of this chapter examines some of the processes involved.

9.3   Crustal accretion at destructive plate margins

9.3.1   *The location of crustal accretion*

Earlier, we noted that continental crustal thicknesses in mountainous regions range up to more than twice the average value of 35 km. Here the distinction between 'Andean' ocean/continent boundaries (Fig. 9.6) and 'Alpine–Himalayan' continent/continent boundaries (Fig. 9.7) will be discussed in terms of their plate tectonic evolution. It will be shown that there is a progressive evolution from Andean-type continental magmatic arcs around some of the margins of a closing ocean (Fig. 9.7, Stage 1), through to Himalayan-type mountains once the ocean is closed (Stage 4 of Fig. 9.7). When continents collide there is little subduction because both are of low density; instead there is considerable uplift. The sediments accumulated on the intervening sea-floor are uplifted, faulted and overthrust. Crustal shortening is very evident in such regions, and igneous activity, though present, is less significant than in the earlier Andean stage.

Magmatic arcs found along active continental margins are characterised by andesite volcanoes, vast linear 'granitic' batholiths parallel to the margin, and variable amounts of clastic sediments. Although these are not 'fold mountains' in the Alpine-Himalayan sense, sometimes extensive volumes of deformed sediments are located on the leading edge of a continental plate where they have been scraped

167

off the subducting ocean lithosphere, as around the North Pacific (Japan, the Aleutians and western USA.). But these sediments do not account for the enormous vertical thickness of continental crust which may be developed, for, in the Andes, they are absent along much of the plate margin. Figure 9.6 shows that the bulk of the upper crust in the active belt comprises a young batholith. High-grade ancient metamorphic rocks of the Brazilian shield pass right underneath the batholith and crop out on the Pacific coast. As Figure 9.4e shows, it is the upper crust, mainly granites overlain by volcanic andesites, that is considerably enlarged. Large-scale *tensional* features, parallel to the plate margin, have been recorded along the length of the Andes – quite opposite to the anticipated effect due to crustal shortening (Myers 1975). The magmas observed at island arcs are similar except that much smaller volumes of intrusive rocks are involved and extrusive rocks predominate (Section 9.3.2). So it seems that the crust thickens at destructive ocean/continent and ocean/ocean boundaries more by *vertical additions* than by the *lateral compression* and shortening, which is more a characteristic of continent/continent boundaries.

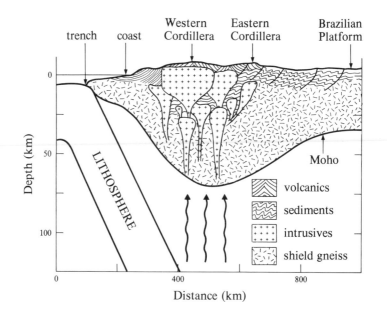

Figure 9.6 Schematic east–west cross section through the central Andean destructive margin to show crustal structure. Arrows indicate direction of magma and volatile streaming from the downgoing plate. Note: 5 × vertical exaggeration. (Source: Brown & Hennessy 1978.)

The essential point of this section is, therefore, that *new material is being added to the continental crust at destructive plate margins*. The accumulated material is due to magmatic processes that originate beneath the continental plate and not to great volumes of continental erosion products. The latter are added back to the continents at their leading edges or in continent–continent collisions, though some may be subducted (Ch. 10) and then become involved in melting and magmatism.

The sites of continental accretion, over active continental margins, usually surround shrinking oceans such as the Pacific (Fig. 8.16). But magmatic activity through subduction can last only as long as the oceans themselves and once a continent/continent suture zone (Alpine–Himalayan) has formed, activity must cease, perhaps to be renewed around some younger ocean. For example, the inevitable collision of Australia with Asia within the next $10^7–10^8$ years will cause activity to cease along the Indonesian arc. Because magmatic arcs are developed around shrinking oceans prior to suturing, it follows that they will flank the fold mountains of continental sutures. Examples of ancient sutures include the Ural mountains, a great north–south chain across central Russia, and the Caledonian mountains across Scandinavia, Britain, Newfoundland and down the Appalachians (a linear mountain chain before the modern Atlantic formed). The eroded remnants of continental plate sutures such as these provide a considerable wealth of information on past crustal movements (Ch. 10).

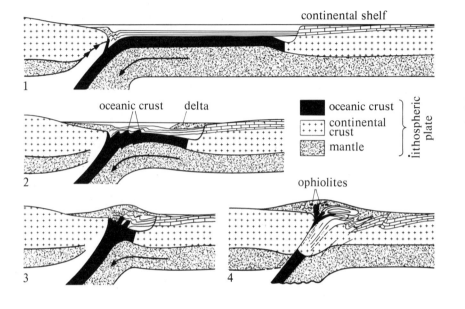

Figure 9.7 Sequence of events (1–4) leading to a continent–continent collision belt with subduction on one margin only. Note that sediments accumulated on the ocean-floor are caught up in the collision together with ophiolitic fragments of oceanic lithosphere. To the left of each section is a continental magmatic arc which is active during stages 1–3. (Adapted from Toksöz 1976. © *Scientific American*, 1975.)

9.3.2 *Magmagenesis and temperatures in the continental lithosphere*

The igneous rocks above subduction zones vary considerably in composition, and in the ratio of volcanic to intrusive material, which decreases with increasing thickening and 'maturity' of the arc. Young island arcs such as the South Sandwich arc, the Marianas and Tongas comprise basalts and basaltic-andesite volcanics. More mature arcs such as those of the West Indies, Central America, New Zealand, Indonesia and Japan (Fig. 9.8) are dominated by calc-alkaline andesites and by dioritic and granodioritic intrusive rocks. Crustal thickness under these arcs approaches normal crustal proportions (up to 35 km) and this is exceeded only in the most mature magmatic arcs, such as the Andes and Rocky Mountains, where intrusive rocks predominate. These, too, are variable, ranging from gabbro to diorite to granodiorite to adamellite, and there is often a correlation between increasing 'acidity' and duration of activity. There is also a correlation between the amount of potassium, and other similar elements, and height above the subduction zone (Fig. 9.8): this is the so-called K–*h* trend.

Although there are many similarities between the volcanic and intrusive suites, prior to about 1970, their origins were divorced by most petrologists. There were several reasons: not the least, it was (incorrectly) thought that most intrusive rocks were granites *sensu stricto* and thus were quite different from the chemical **169**

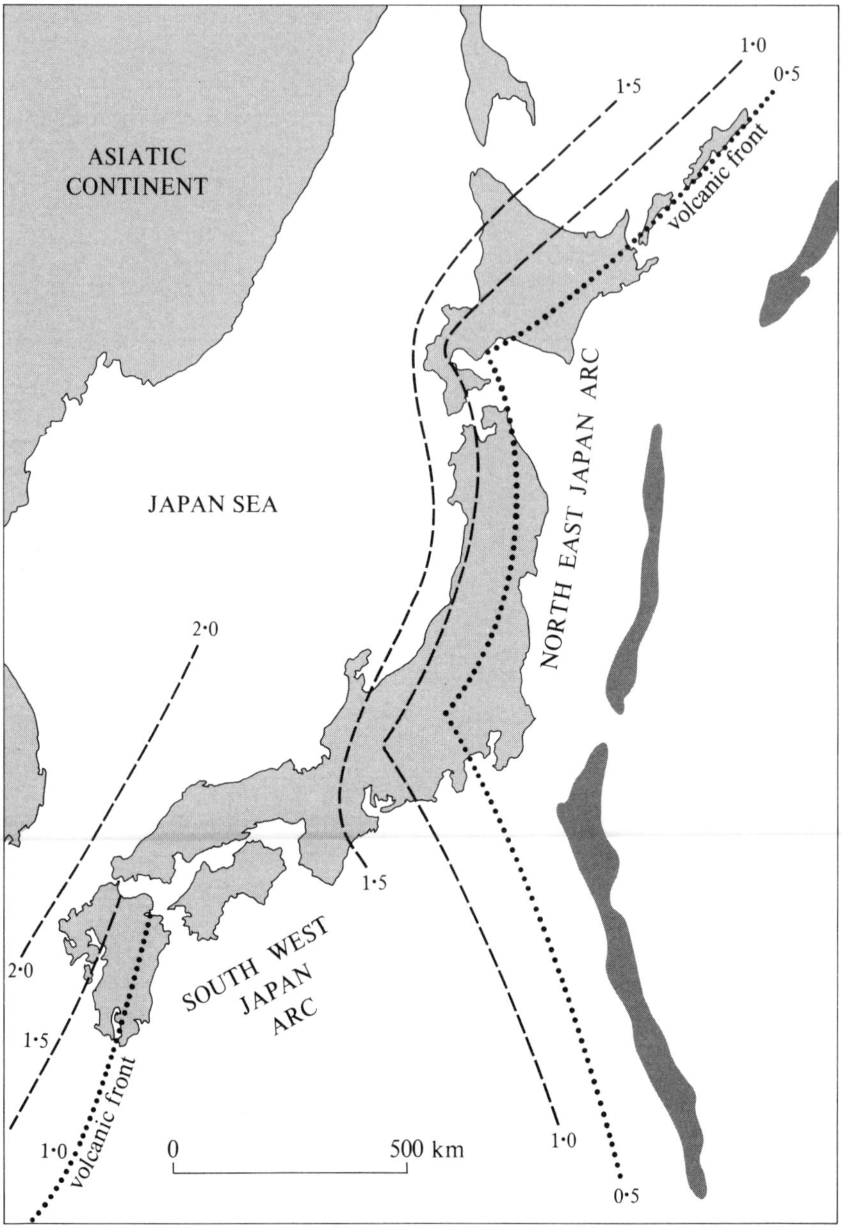

Figure 9.8 The Japanese trench (shaded) and island arc showing the positions of the volcanic fronts and contours of the increasing K_2O content of lavas across the arc (standardised for lavas with 55% SiO_2). Subduction of Pacific Ocean lithosphere is in an east–west direction; the depth to the subduction zone increases beneath Japan in that direction. (Data from Dickinson 1968, Miyashiro 1972.)

composition of andesites. Secondly, andesites abound in island arcs where, in many cases, there are few granites and little continental crust. Andesite melts are therefore generated by melting in the mantle and this was contrary to the belief that granite melts could form only in the continental crust. This followed partly from the experimental work of Tuttle and Bowen (1958) which showed that granitic melts can be produced from average crustal metasediments provided that the pressure–temperature conditions of the 'beginning of melting' curve pertain (Fig. 9.5). Nowadays, it is thought that temperatures in the thick sediment piles of closing oceans and marginal basins (behind arcs) are rarely adequate and this raises the question of whether the intrusive rocks of mature arcs, like the extrusive rocks they

170

resemble chemically, are generated beneath the crust. The following boxed section of the text describes how the relevant crustal temperatures are estimated.

Temperatures in the continental crust. Table 8.2 shows that half the average surface heat flow of the continents is produced from within the crust, due to its high content of radioactive isotopes. However, this heat production is not uniformly distributed but appears to decrease logarithmically with depth (Lachenbruch 1970). This was deduced from the observation that heat flow (q) correlates with surface heat productivity (A) in most continental regions (Fig. 9.9). Regions with the lowest heat production are usually the oldest, Precambrian shield areas from which the greatest thickness of material has been removed by erosion. The straight line in Figure 9.9 has the equation

$$q = q^* + A_0 h$$

where q^* is the heat flow from beneath the radioactive layer whose surface heat production is A_0 and whose thickness is a function of h (Birch *et al.* 1968). Lachenbruch (1970) found that, to maintain this linear relationship during erosion, the heat produc-

tivity A_z at any depth z must be described by $A_0 \exp(-z/h)$, hence the logarithmic decrease of heat production with depth mentioned above. The origin of this distribution is likely to be related to the upward migration of magmas containing the strongly lithophilic elements K, U and Th (Brown & Hennessy 1978) thus leaving the lower crust relatively depleted (Section 9.2.3). Because heat flow decreases with depth, the geothermal gradient is most steep near the surface (in a conductive layer, the geothermal gradient is closely proportional to heat flow – Note 9). Hence, the shape of the geotherm through the continental tectosphere must be curved and it is not valid to extrapolate near-surface geothermal gradients, measured in boreholes, down to deep crustal levels. The geotherm shown in Figure 9.10 is for the top 100 km of the continental tectosphere, and is drawn on the assumption that the likely temperature at the 100 km depth to the asthenosphere approaches 1200°C. From this geotherm, the temperature at any particular depth may be deduced.

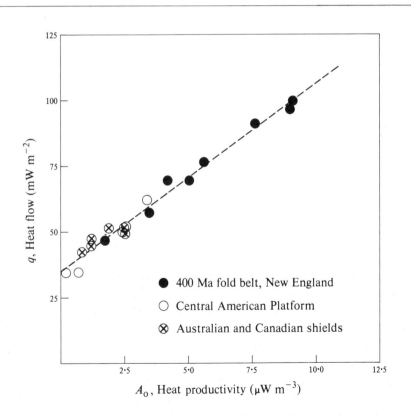

Figure 9.9 The correlation of heat flow and heat productivity for various continental areas. (Source: Sclater & Franchetau 1970. Used with the permission of the Royal Astronomical Society.)

171

Figure 9.10 shows the variation of temperature through a typical piece of stable continental crust, 35 km thick, and down to 100 km depth. The continental geotherm has a gradient close to $30°C\,km^{-1}$ near the surface but decreases to less than $5°C\,km^{-1}$ at 100 km depth. Also shown are the wet and dry melting curves for granite (see Fig. 9.5 for the former): temperatures in normal crust barely intersect the (shaded) field of granite melting. Since lower crustal rocks are also thought to be mainly dry (Section 9.2.2), it is unlikely that granite melting will occur spontaneously. However, a rising hot body of magma from mantle depths can perturb the geotherm towards higher temperatures, perhaps causing partial melting and an increase in the volume of the rising magma body. So, a prerequisite for crustal melting is that magmas must already be rising from subcrustal levels and this suggests a *common* source for both intrusive and extrusive magmas. These comments apply to most destructive margins but in continent–continent collision zones, crustal temperatures may be increased to melting conditions by frictional effects and abnormal radiogenic heating where sedimentary layers, rich in heat-producing elements, are thrust upon each other (Oxburgh & Turcotte 1974). This points to crustal recycling at Alpine/Himalayan boundaries and crustal growth at island and continental arcs.

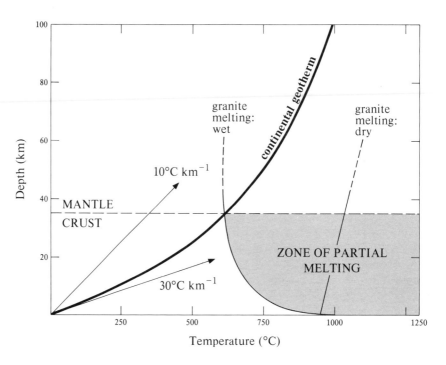

Figure 9.10 A typical continental geotherm extrapolated, with decreasing geothermal gradient, to a temperature of $c.1000°C$ at the 100 km depth of the asthenosphere. Wet and dry granite melting curves (after Brown & Fyfe 1972) are shown for comparison. Shading indicates the zone in which partial melts of granite can be generated.

Next, the conditions necessary for melting beneath the crust in a subduction environment will be considered. The subducted material comprises an upper 5–10 km of mainly basaltic crust, overlying depleted peridotite. The latter is unlikely to melt during descent back into the mantle, because it is already the residuum after partial melting at the ocean ridge. But what about the ocean crust? It is characteristically hydrothermally altered by interaction with sea water and so it contains many hydrous minerals such as zeolites and amphiboles. As temperatures rise, during

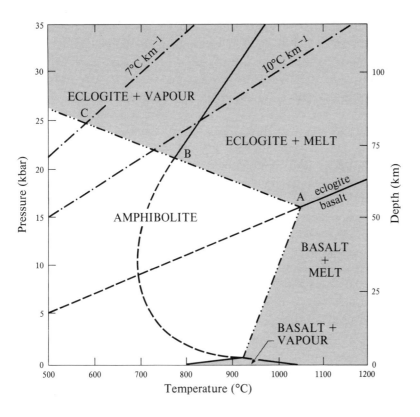

Figure 9.11 Stability limits for amphibole in subduction zones. Breakdown of amphibole occurs in the shaded part of the diagram. Also shown are the wet melting curve for basalt (curved) and the basalt/eclogite phase boundary. (Source: Fyfe & Brown 1972.)

subduction, metamorphism will ensure that most of the basalt is converted to amphibolite. In Figure 9.11, amphibolite is stable in the unshaded area, but breaks down to release water in the shaded area. There is a change of slope on the line between these two areas where it crosses the basalt/eclogite phase boundary (cf. Fig. 7.2) and, in plunging a cold oceanic plate into the mantle, only the high-pressure, low-temperature part of the diagram – above this boundary – is relevant. The downgoing plate is heated by conduction from the surrounding molten mantle and by friction at its boundaries giving a geothermal gradient down the subduction zone of about $10°C km^{-1}$ (Fig. 9.11). The *curved* line is the wet melting curve for basalt–andesite compositions and indicates where melting will start if free water is available; that is, water not locked up in the silicate structure of amphibole. Therefore, only the *solid* part of this line is relevant to melting, outside the field of amphibole stability. As the oceanic lithosphere is plunged down into the mantle, there are two possible reactions.

(a) Across AB:

> amphibolite → melt + depleted eclogitic residue

The melt produced is in the compositional range of andesite to basaltic-andesite.

(b) Across BC, at lower thermal gradients:

> amphibolite → vapour + quartz eclogite

173

In this case, the water vapour penetrates the overlying hotter mantle, carrying with it abundant mobile chemical species (mainly lithophilic elements) because of the solubility of many silicate minerals in hot water. Because the overlying mantle is relatively hot and dry, the effect of adding volatiles is to depress the melting temperatures there and to cause partial melting.

The subducted plate can therefore initiate melting in two ways: by generating either a partial melt from its own substance or the volatiles capable of producing a melt within the adjacent mantle. In either case, the bulk of the subducting plate continues its descent as refractory eclogite and is absorbed into the mantle at greater depths. But the melt phase, being of relatively low density and therefore buoyant, rises to invade the crust or form a volcanic arc (Fig. 9.13, p. 177).

On rising into the crust, the magma may be hot enough (about 1000°C) to melt some of the surrounding crust and become more acidic. Alternatively, as it cools, it may fractionally crystallise first-formed calcic feldspar and pyroxene crystals, so that the residual liquids become more acidic before their final crystallisation. Whereas in young island arcs, with thin crust, all the magma may be erupted, if there is a considerable thickness of continental crust much of the magma may cool to its freezing point *before* reaching the surface, and so form intrusions. The tear-drop shaped plutons shown schematically in Figure 9.6 are based on (i) observations of different erosion levels in batholiths (e.g. Bateman & Eaton 1967), and (ii) on modelling experiments with immiscible fluids. Collectively, many plutons make a batholith.

Because of the melting processes just described, the mantle above subduction zones is depleted of any low-temperature-melting components by the ascent of magma streams during the active history of the arc. If, in general terms, the continental masses have accreted from the centre outwards by the addition of materials at arcs around their margins (Ch. 10), then the refractory mantle beneath them will have grown outwards at the same rate. This provides an explanation for the generation of the mantle part of the continental tectosphere which moves around with the continents, rather like the submerged part of an iceberg (Sections 8.4–8.6).

9.3.3 *Magma variations in space and time: a genetic model*

Evidently, the production of igneous rocks is a major destructive plate margin process and in this section we look at some of the end products – the igneous rocks exposed at the surface – in order to discover how they relate to their parental magmas. Much of the evidence is taken from detailed work on two batholiths formed in the last 100 Ma: the Sierra Nevada batholith of California, studied by Bateman and Dodge (1970), and Hamilton and Myers (1967), and the Peru–Chile coastal batholith studied by Pitcher (1978) and others. Their most important characteristics are summarised in Table 9.3. Both earlier and later batholiths have been recognised in the western Americas, but at different distances from the trench. This means that the critical dehydration reactions which initiate and focus magmatism must change their inland distance from the trench between cycles of activity, possibly due to periodic changes of subduction angle. Thus, for example, a recent development in Peru has been the shift of activity some 80 km east from the coastal batholith, to form a new batholith.

Table 9.3 Some characteristics of continental arc magmatism.

(1) repeated magmatic cycles, each lasting 75–125 Ma

(2) progression with time from intermediate to acid magmas, both within individual and between successive cycles

(3) associated with (2), Rb/Sr and K/Na ratios increase with time

(4) the ratios in (3) also increase with distance from the trench, e.g. the K–*h* trend in Figure 9.8

(5) $^{87}Sr/^{86}Sr$ initial ratios are low and sometimes increase with time and distance from the trench (Fig. 9.12)

(6) $^{87}Sr/^{86}Sr$ ratios of basement rocks exceed magma initial ratios at the time of formation

(7) volcanic and intrusive magma trends are often inseparable

The geochemical trends 2–4 in Table 9.3 were originally proved for volcanic rocks (Fig. 9.8, for example) and later extended to intrusive rocks. Local changes between such extremes as early gabbros and late adamellites have been recorded within an 80 Ma time span. Isotopic data became widely used as indicators of magmagenesis and evolution during the 1970s. Trends in initial strontium isotope ratios are noted as items 5 and 6 of Table 9.3. This ratio discriminates between magmas derived by melting of mature crust and those of mantle origin (see Note 5). Figure 9.12 shows a good correlation between initial ratio and age, and the younger, more acid rocks have the higher initial ratios. Moreover, the trends of intrusive and extrusive rock initial ratios are overlapping (item 7, Table 9.3), again suggesting a common parentage.

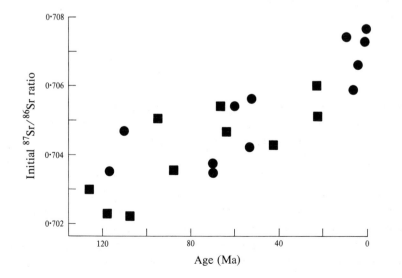

Figure 9.12 The trend of increasing strontium isotope initial ratios for intrusive (squares) and extrusive (circles) rocks from northern Chile. (Source: McNutt *et al.*, 1975.Used with the permission of the Elsevier Scientific Publishing Co.)

What else can be learnt from these values? The initial ratios of igneous rocks are usually the same as the ratio in their source region at the time of formation (Note 5). If modern arc magmas are generated by melting of ocean crust or adjacent mantle, then their ratios should be similar to those in these source rocks. The present-day mean ratio for ocean tholeiites, generated at the ridge zones, is 0.7028 and that for ocean island alkali basalts is 0.7039 (see Section 7.2.2 for definitions of these rock types). If the Earth's crust and mantle are taken together, they have an Rb/Sr ratio of about 0.03 and, after 4600 Ma, this means that the initial ratio of the Earth would evolve to 0.705. However, because the Rb/Sr ratio of crust-forming magmas is much higher than 0.03 (Table 9.2), the mantle now has a rather lower Rb/Sr ratio and hence provides magmas with lower $^{87}Sr/^{86}Sr$ ratios. (The two types of oceanic basalt have different initial ratios, which implies discrete and unmixed source regions in the shallow mantle (Fig. 9.13). This heterogeneity in the mantle is not yet fully understood, but argues for a great deal of complexity in mantle convection for the two regions to preserve their geochemical identity – see O'Nions *et al.* 1978). So basalts of the ocean crust *should* carry ratios in the range of 0.702–0.704 from the ridge to the subduction zone. But many of them become hydrothermally metamorphosed by sea water which has an $^{87}Sr/^{86}Sr$ ratio of 0.707 at the present day (see Fig. 10.16). Some sea water strontium may therefore exhange with ocean crustal strontium, and this means that a range of $^{87}Sr/^{86}Sr$ up to 0.707 will, in theory, be available in the subduction zone.

The earliest initial ratios shown in Figure 9.12 lie towards the bottom end of this range and, therefore, the early melts have simple, unmodified ocean crust or upper mantle strontium isotope initial ratios. But what of the later stages? Here initial strontium ratios rise above accepted mantle values but still do not record a strong continental crustal imprint. In the Central Andean case (Fig. 9.6), any fusible Precambrian basement gneisses will have developed crustal ratios in the range 0.710–0.780, depending on their age and Rb/Sr ratio (Fig. N5.2). A mixed source is implied whereby the supply of low ratio mantle-derived magmas has been enriched by either seawater strontium or crustal strontium. The latter may be the more likely explanation for the increase in initial ratios with time in the Central Andes (Fig. 9.12 – see also Brown 1979). As time progresses, and the crust is thickened by magmatism, it will also heat up so that temperatures locally may considerably exceed the average crustal geotherm of Figure 9.10. Small pockets of crustal melt will be generated with high initial strontium ratios, and will be available for scavenging or assimilation as the active cycle proceeds. Such melts are likely to be granite (*sensu stricto*) and, if these are progressively added to or mixed with mantle-derived intermediate magmas, the end-product with time will become more acidic. Thus, the trends identified in Table 9.3 are all predictable in terms of a mantle magma source which is perpetuated throughout the cycle whose products mix with small, but increasing amounts of crustal melt as activity proceeds. This process ensures that the lower crust, from which the melts have been extracted, will become refractory granulite.

Crustal thickness and the presence of ancient basement both seem to be important controls over these acidification and initial ratio trends because they become more obvious in the most mature arcs founded on ancient basement (see Armstrong *et al.* 1977, Thorpe & Francis 1979). For example, the basaltic andesites of Central America, Colombia, Ecuador and southern Chile are good examples of magmas which penetrated only 20–30 km of crust rather than the 50–70 km where crustal scavenging and acidification trends are recognised. Recycling of materials by magmas derived from within the crust is therefore most important in the most mature continental arcs; elsewhere, almost all the igneous material produced over destructive margins constitutes new additions to the continental crust. Figure 9.13 summarises the two-stage process which is needed for continental crustal accretion from the mantle: partial melting to create ocean crust at constructive margins, followed by further partial melting at subduction zones.

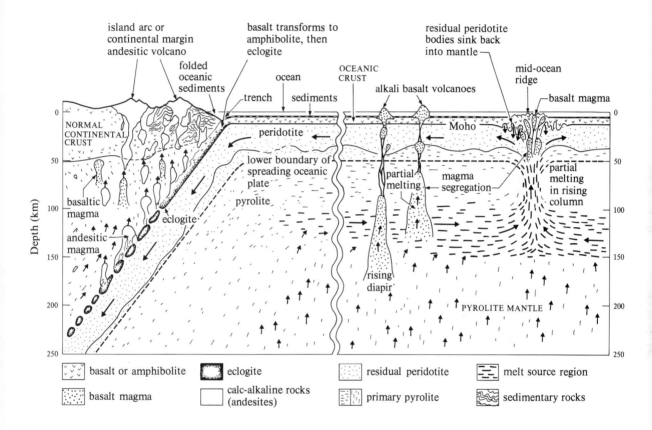

Figure 9.13 A petrological model of melting events in the mantle beneath ocean ridges, ocean islands and continental/island arcs. (Source: Ringwood 1974. Used with the permission of the Geological Society of London.)

9.4   A quantitative estimate of crustal accretion

We have shown that most additions to the continental crust occur at destructive margins of both ocean/continent and ocean/ocean types. The net effect is either to create new crust at island arcs which, one day, will be accreted to a leading continental plate edge, or to thicken existing continental crust. Contemporary observations of these zones of crustal activity allow the rate of accretion to be estimated.

Crustal thickening occurs to the extent of between 10 and 30 km during a 100 Ma active cycle, and a value of 20 km average is used in the following calculation. A characteristic width over which the crust is active during one cycle at a destructive margin is, perhaps, 60 km. The third dimension required is the total width of active

177

destructive margins at the Earth's surface which is 40 000 km. This gives a total volume of $20 \times 60 \times 40\,000 = 4.8 \times 10^7\,km^3$ of continental crust which will be accreted during 100 Ma of activity along destructive plate margins, a rate of almost $0.5\,km^3$ per year.

It is useful to compare this value with the volume of existing continental crust and the time available to produce this crust. The total area of all continental crust is about $150 \times 10^6\,km^2$ and, taking an average thickness of 35 km, this gives $5.25 \times 10^9\,km^3$. Thus, at the modern rate of crustal accretion, slightly more than 10 000 Ma would be required to produce the continents – more than twice the time available! Yet our accretion rate is likely to be an overestimate, which would lead to an underestimated time, because continental erosion has not been considered. Most of the erosion products wind up as sea-floor sediments: some form a fairly permanent cover to the continental shelf–platform areas, but others are deposited in the deep ocean. Here, they are carried along the sea-floor spreading conveyor system as layer 1 of the ocean crust until they become involved in continental margin processes. In some areas, the sediments pile up along the margin, once more becoming part of the continent, as along the western coast of North America, eastern Japan and elsewhere (Fig. 10.12). To the west of the Andes, no such sedimentary prism has been identified, probably because the present drainage system carries most of the erosion products from South America into the Atlantic rather than the Pacific. In the case of a closing ocean (Fig. 9.7), the sediments are uplifted and returned to the continental crust. Despite all these considerations, it may be that part of the sedimentary layer 1 of ocean crust is *subducted* and, if so, this material will be available for remelting as new continental crust forms again. If some of the new arc magmas are recycled, they cannot be classed as new crustal material from the growth point of view. However, recycling cannot proceed indefinitely, or contribute in any large measure to continental and island arc magmatism, for then much higher initial strontium isotope ratios would be recognised. The accretion rate of $0.5\,km^3\,a^{-1}$ given above may, therefore, be an overestimate; certainly, it is not an underestimate. The question of sediment recycling on a large scale is re-evaluated in Chapter 10 in relation to the pattern of crustal growth.

To account for the sheer volume of continental crust which does exist, only one possibility remains: that of *a much higher accretion rate at some time or times in the Earth's past history*, and this is the theme on which we open the last chapter of this book.

Summary

1   The Earth's continental crust comprises: (i) ancient Precambrian stable shield areas, (ii) flat-lying platforms with a cover of gently folded sediments overlying more shield (or cystalline basement), and (iii) younger mountain belts. The latter include continental magmatic arcs over ocean/continent destructive boundaries and fold mountain belts which mark the line of continent: continent collisions. The average thickness of the continental crust is 35 km but, to achieve isostatic balance, mountainous regions range up to 90 km thick.

2   The continental crust comprises two major layers beneath a variable thickness

of metamorphic and sedimentary rock cover. On average, the upper crust is granodiorite, both in terms of rock type and chemistry. The lower crust comprises mainly high-temperature, refractory granulites whose chemical composition, apart from being depleted in lithophilic (and heat-producing) elements, is dioritic.

3   The major igneous rock types of continental arcs are extrusive andesites and, more important, diorite–granodiorite intrusive suites which bear a strong chemical resemblance to each other. Apart from quite small contributions by crustal melting, mainly in the thickest continental arcs, these magmas are derived from dehydration or melting of the downgoing slab of oceanic lithophere and the overlying mantle because:

(a)   crustal temperatures are insufficiently high for major magmagenesis in the top 35 km (except in continent/continent fold mountains);
(b)   strontium isotope initial ratios of all the igneous rocks involved are strongly influenced by the mantle; and
(c)   there is little lateral thickening of continental arcs so that vertical thickening from subcrustal sources is necessary.

4   A model is suggested whereby continental growth and the development of thick (400 km) refractory, subcontinental tectosphere roots are linked by the partial melting process above subduction zones. As new crust is accreted, principally around continental margins, the requisite magmas stream up through this mantle region depleting it of its melt fraction. Therefore, growth of the continent and its depleted source region progress outwards and result from the same sequence of events in space and time.

5   New continental crust accretes at the present day through magmatism at island and continental arcs. The modern rate of continental accretion is estimated at about $0.5 \, km^3 a^{-1}$, which is less than half the rate needed over the Earth's history to 'grow' all the continental crust.

Further reading

General books:
  Wyllie (1976): crustal geophysics and geochemistry.
  P. J. Smith (1973): crustal geophysics, seismic structure, etc.
  Turekian (1972); Mason (1966): reviews of the composition of the crust.
  Gass (1979): magmagenesis.
Advanced journals:
  Ronov & Yaroshevsky (1969): composition of the crust.
  Sclater & Franchetau (1970); Davies (1979): crustal temperatures.
  Brown & Hennessy (1978): crustal temperatures and granite magma genesis.
  Ringwood (1974): evolution of island arcs.
  Pitcher (1978): description of the Peru coastal batholith.
Advanced book:
  Thorpe (1981): reviews of magmatic arcs, their magmas and evolution.

10 Evolution of the Earth's continental crust

10.1 The framework of crustal evolution

In Section 9.3 we explained how the Earth's crust evolves by a two-stage fractionation process, first at spreading ridges, then at subduction zones. If the principle of uniformitarianism – i.e. that the processes observed today are similar to those of the past – is correct, why is it necessary to discuss the topic again? The reason is that present-day processes might not have operated in the remote past, primarily because of the much greater heat production by radioactive elements. Apart from early peaks due to gravitational, tidal and short-lived radiogenic heating (e.g. ^{26}Al, ^{244}Pu in Fig. 10.1), the Earth's internal heat production has depended on the four long-lived radioisotopes shown in Figure 10.1. The rate of internal and surface tectonic processes probably has decreased in proportion to this decreasing heat production and this may explain why the postulated modern rate of continental accretion in Section 9.4, i.e. $0.5\,km^3\,a^{-1}$, is less than that needed to produce the entire mass of the continental crust in 4600 Ma. Although most workers agree about the Earth's long-term thermal decay (dashed line in Fig. 10.1) there is far less agreement about the *mechanism* of accretion throughout geological time and about the *past rate of continental growth*. The term **crustal accretion** has, so far, been used to indicate just the addition of new material to the continents at arc-type margins without taking account of material recycled or removed by erosion, etc. It is, therefore, not necessarily the same as 'crustal growth' which implies the net addition of material to the continents.

To explain, during the late 1970s there were two schools of thought on continental growth (Fig. 10.2):

Figure 10.1 Heat production from the important radioactive isotopes incorporated at the Earth's formation. K, U and Th values have been extrapolated back from present-day average abundances in the crust and mantle, while Al and Pu values are estimated from nucleosynthesis theory using stable isotopes in the Allende meteorite. (After O'Nions *et al.* 1978. Used with the permission of the Royal Society.)

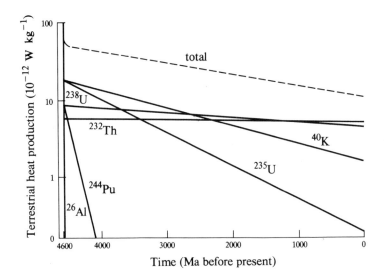

(a) One argued that only small amounts of continent were produced by the initial heating of the Earth and that the growth of continents has been continuous and proportional to the heat output of long-lived radioactivities (compare curve 1 of Fig. 10.2 with Fig. 10.1). This model assumes that all continental erosion products are added back to the continents along active continental margins and in continent–continent collisions (Fig. 9.7). So the continental accretion rate calculated earlier is equal to the growth rate (for reviews, see Windley 1977, Moorbath 1978, Brown 1979).

(b) The other proposal was that continental masses were established during the first 2000 Ma with little subsequent change in volume during most of geological time. Here, the material accreted to the continents is balanced by the subduction of continental erosion products back into the mantle. Although there is accretion, there is not net growth: curve 2a in Figure 10.2 (see, also, Hargraves 1976, Fyfe 1978). The main difference between this view and model 1 is that the crust is still growing according to the latter.

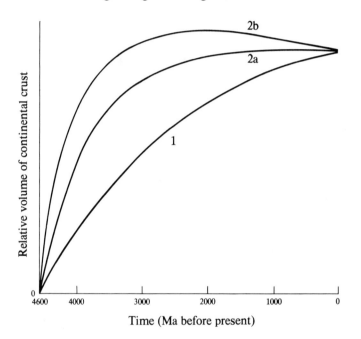

Fyfe (1978, 1979) modified model 2 as a result of estimating the rate of sediment accumulation in ocean basins, which gives the maximum possible rate at which sediment could be subducted. His estimate was $2 \, km^3 \, a^{-1}$, which is four times that of accretion at modern magmatic arcs. He used this to suggest that the continental volume could actually be *shrinking* (curve 2b, Fig. 10.2).

It is difficult to make a final choice between these models (Section 10.5), but a variety of additional evidence bears on the past rates of crustal accretion and growth (Sections 10.2–10.4). The key to past processes can be found only in the continental crust because the only exposed Precambrian rocks (> 570 Ma age) are continental, often occurring in continental nuclei, or cratons (Fig. 9.2) which comprise about 15% of the Earth's surface. The *Archaean* period (> 2500 Ma age) is recorded only in even smaller areas because most Archaean crust has been obliterated by long-continuing geological processes. However, there are many common features in these small, widely separated Archaean areas. They all comprise areas of high-grade metamorphic **gneisses** surrounded by low-grade volcano–sedimentary sequences – the so-called **greenstone** belts (Fig. 10.3). The oldest known Archaean rocks from Greenland, Zimbabwe and North America are about 3500–3800 Ma in age. This raises another fundamental question: What happened during the interval between the Earth's origin at 4600 Ma ago and the formation of the oldest crustal rocks at 3800 Ma? (Section 10.2).

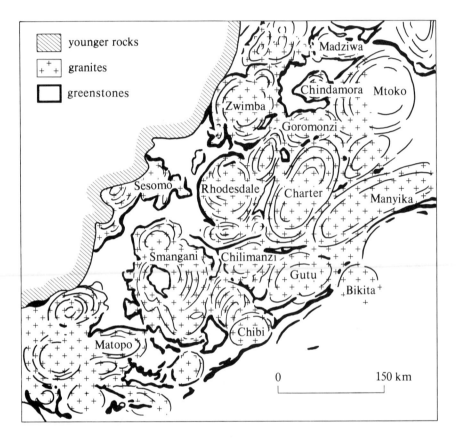

Figure 10.3 Sketch map of the Archaean granite–greenstone pattern of the Zimbabwe craton. The position of the Zimbabwe craton is shown in Figure 10.8. (After McGregor 1951.)

The later Precambrian or *Proterozoic* period (2500–570 Ma ago) was characterised by markedly different geological and tectonic conditions. The gneiss–greenstone structures, which are small-scale (from 10 to 100 km across) became trapped within large, stable continental platforms with lateral dimensions up to thousands of kilometres. Extensive sedimentary sequences covered an older crystalline rock basement and linear **mobile belts** developed both within the stable blocks and around their margins. These were the sites of intense deformation and igneous activity (see Fig. 10.4). Evidence for modern plate tectonic processes can be traced back into the Proterozoic and may be responsible for some of the craton–mobile belt configurations. For Archaean times, the term **permobile** (rapid motion) is often used to describe the intensity and dynamism of tectonic mechanisms which, again, may have been plate-like in operation. Both the scale and intensity of tectonic processes have changed and strict uniformitarian concepts cannot be applied to earlier geological times. The main aim of this chapter is to understand the history of the crust and the history of surface conditions on our planet under the following age categories:

Archaean	pre-2500 Ma	[Gr. *arkhios*, ancient]
Proterozoic	2500–570 Ma	[Gr. *proteros*, former; *zoos*, living]
Phanerozoic	post-570 Ma	[Gr. *phaneros*, visible; *zoos*, living]

Finally, we present an evolutionary model for the crust and the Earth as a whole. (Note that the diagram included in the rear flap provides a summary of the main points in this chapter.)

Figure 10.4 Archaean granite–greenstone terrain in the East Pilbara region of Western Australia. Light areas are granite gneiss domes and the finely striped regions between are greenstones. Archaean rocks are surrounded by Proterozoic sediments and volcanics (dark areas), while a cover of young Tertiary sediments appears in the south-west and north-east corners of this plate. (Source: ERTS image, No. 1148–01282, December 1972. Scale, 1:1000 000.)

10.2 The Archaean

10.2.1 *Geology and tectonics*

Most well preserved Archaean terrains, in the cores of ancient shield areas, show the characteristic high-grade gneiss plus low-grade volcano–sedimentary greenstone patterns. Figure 10.3 is a map of the Zimbabwe craton which comprises subcircular 183

granitic plutons separated by deformed, schistose greenstones. Both the granites
and greenstones formed over a range of times from about 3500 Ma to about
2600 Ma (Wilson *et al.* 1978). Isotopic ages from other cratons containing similar
high-grade rocks vary from late Proterozoic (about 2300 Ma) to the oldest known
rocks from Greenland, the Amîtsoq gneisses (3800 Ma). Moorbath (1977) recognised clustering of ages in the Archaean which may signify episodic crustal growth
(Section 10.5.1).

The low-grade greenstone sequences vary considerably in thickness, composition
and degree of metamorphism. Much of the material was volcanic in origin, but it
includes clastic sediments (immature molasse and greywackes) of andesite–dacite
composition. Some basic, and even ultrabasic, lavas (typically magnesium-rich
komatiites) are also found. The linear greenstone belts we see today range up to tens
of kilometres across and hundreds of kilometres long; they may have been narrow,
elongate depositional basins that have undergone small amounts of deformation
and metamorphism. Also associated with greenstones, there is often a group of
cyclic sedimentary deposits comprising deep-water shales, shallow-water immature
sandstones and *banded-ironstone* and chert (fine-grained silica) *formations*, or
BIFs. Although the relative amounts of high- and low-grade material exposed in
Archaean cratons varies, there is still a striking uniformity of composition and
major rock units from such widely separated regions as South Africa, Zimbabwe,
Canada, Western Australia, Greenland and India (see Windley 1977).

Evidently, the Archaean crust had formed by 3800 Ma ago, but there is no record
of crustal events before this time, except that broken-up remnants of greenstone-like
materials occur within the Amîtsoq gneisses. It is not known whether the Earth
formed an early anorthositic crust, like that of the Moon, but there is other evidence
for higher temperatures than today from preserved Archaean rocks. Mg-rich
komatiite lavas require 1700°C temperatures for their formation since they are
produced by near total melting of mantle peridotite. Moreover, many komatiites
have similar ages to normal basalts (about 1100–1200°C formation), suggesting
rather variable temperature distributions in the source region. This may reflect the
higher heat production of the early Earth (Fig. 10.1) which may also have led to
higher thermal gradients in the mantle. As shown in Section 8.6.1, the Rayleigh
number increases with thermal gradient, leading to greater irregularity in convection patterns and to more erratic temperature distributions. Other responses to the
higher Archaean heat production may have been more vigorous mantle convection
and a greater number of convection cells than today (Bickle 1978). Possibly,
therefore, regions of magma penetration through the solidified skin, or lithosphere,
of the Earth were more closely spaced, and small mobile fragments of lithosphere
acquired greater velocities than today.

Although the greater heat production and vigour of mantle convection in the
Archaean is not disputed, a much debated consequence is that the isothermal
boundary separating the yielding asthenosphere and rigid lithosphere *may* have
been closer to the Earth's surface. Average present-day thermal gradients across the
continental lithosphere are around 10°C km^{-1} (Fig. 9.10) but gradients approaching 100°C km^{-1} have been suggested (Fyfe 1974) for the Archaean. If the
temperature at the base of the lithosphere then, as now, was about 1200°C, it follows
that the lithosphere might have been only 10–15 km thick. This led to the suggestion

that the early lithosphere (i.e. pre-3800 Ma) was unstable, fragmented and became resorbed into the mantle (Hargraves 1976) which conveniently explains the lack of crustal relics older than 3800 Ma. As the initial vigour of mantle convection waned and the lithosphere thickened, Hargraves envisaged that stable crustal fragments swept together to form large cratons. A further consequence of the thin lithosphere model is that the continents had low relief and that the entire globe was covered by water from vigorous volcanic outgassing following the Earth's accretion (Fryer *et al.* 1979).

The thin Archaean lithosphere model has been countered by the recognition of high-pressure (7–10 kbar) metamorphic minerals and granulite facies rocks in eroded Archaean crust. Thus Windley (1980) and Watson (1978) have argued that continental *thicknesses* locally reached modern dimensions and this is possible if temperature gradients had marked lateral variation. Seismic data (e.g. Fig. 9.4a) indicate that present crustal thicknesses beneath high-grade Archaean outcrops are often 35–40 km. Therefore, some continental crust must have emerged from the Archaean oceans and this is also consistent with the sedimentological evidence for 'massive erosion of extensive uplifted continental highlands before deposition of late Archaean sediments' (Windley 1980, in discussing the Indian Archaean). In Figure 10.5 these ideas are developed into a mobile micro-continent model for Archaean crustal processes in which the emergent domes, or arcs, comprise the meta-morphosed high-grade gneisses. The crust was extended by many convection currents in the mantle which developed subduction zones beneath the micro-continents, and locally the crust was thinned to form back-arc extensional basins. Into these basins were erupted, first, high-temperature basalt–komatiite lavas and,

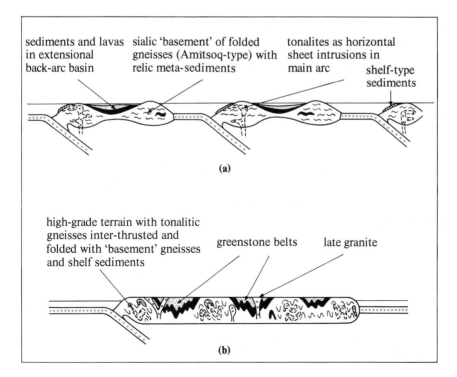

sediments and lavas in extensional back-arc basin

sialic 'basement' of folded gneisses (Amîtsoq-type) with relic meta-sediments

tonalites as horizontal sheet intrusions in main arc

shelf-type sediments

(a)

high-grade terrain with tonalitic gneisses inter-thrusted and folded with 'basement' gneisses and shelf sediments

greenstone belts

late granite

(b)

Figure 10.5 Mobile micro-continent model for the genesis of Archaean granites and greenstones. Diagram (a) represents the mobile stage where folded gneisses and island-arc magmas form the domes. Volcanic and sedimentary proto-greenstone materials accumulate in back-arc basins. In (b) the micro-continents have aggregated to form the structures now observed. (After Windley 1977. Used with the permission of John Wiley & Sons Inc.)

185

later, andesite–dacite lavas, similar to those of modern arcs (Tarney *et al.* 1976). The lavas became buried by clastic sediments, derived from the emergent domes, and the final evolutionary stage was the aggregation of the micro-continents which closed up the back-arc basins and produced the characteristic greenstone belt structures (cf. Figs 10.3, 10.4 and 10.5b). This model of greenstone belt formation is more widely accepted than other models which regard greenstones as vestiges of oceanic lithosphere trapped in front of the domes, like ophiolites. One of the attractions of this permobile model is that the concept of micro-continents (perhaps about 100 km wide and about 30 km thick), each with its own subduction zone, is consistent with vigorous mantle convection and laterally variable temperatures. The highly mobile continents would thicken easily by energetic collisions leading to considerable deformation and metamorphism. The 'permobile' model, with thick continental crust, may imply more coupling between the lithosphere and convecting mantle beneath by basal drive forces than appears to be the case today (Section 8.8).

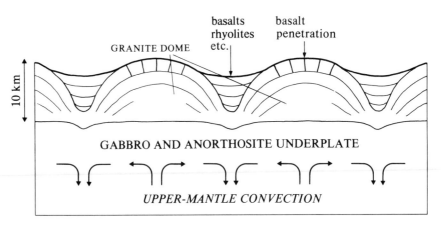

Figure 10.6 Model for the evolution of Archaean granite-greenstone belts with limited lateral movement and vertical tectonics. See text for further explanation. (Modified from Fyfe 1974.)

An alternative explanation of Archaean crustal patterns, with relatively static lithosphere and more uniformly high temperature gradients (Fyfe 1974, 1978; Fig. 10.6) necessitates a thin lithosphere (as described earlier). This lithosphere comprised mainly a crustal layer, of basaltic composition, and above the upwelling limbs of mantle convection cells, granitic domes were developed. Here granite magmas, generated by crustal partial melting, rose to give the complex structures now found in the eroded domes. Between the domes, the crust was thinned, cool material was sucked back into the mantle where it was heated, partially melted and then erupted to form greenstone lavas. The crust was thickened by the underplating of more mantle fusion products, probably gabbros but also including anorthosites formed by low-pressure crystallisation of mantle melts (Section 5.4.2). This relatively *static* model of the Archaean crust – vertical movements but little horizontal motion – has been termed 'hot spot tectonics' (Fyfe 1978).

To summarise, the two models for Archaean crustal tectonic patterns are: (i) the permobile model, where the lithosphere is locally thick, there are land areas available for erosion and sediment formation, the micro-continents collide and aggregate, and greenstones form in extensional back-arc marginal basins; and (ii) the hot spot model, where the lithosphere is thin, there are few (if any) land masses, no lateral plate movements, and greenstones form over descending convection be discussed in Section 10.5.

10.2.2   *Archaean atmospheres and primitive life*

The Earth's present atmosphere is due primarily to volcanic outgassing. Any gaseous envelope which condensed around the Earth when it first formed has been either extremely diluted, or totally replaced, because the noble gases are very strongly depleted compared with their cosmic abundances (Fig. 4.12; Fisher 1976). Apart from oxygen, the major gases found in the atmosphere (N_2, 78%; O_2, 21%; Ar, 0.93%, CO_2, 0.03%, plus variable amounts of H_2O) are all found in volcanic gases. The ratio of argon to nitrogen in volcanic gases is the same as in the atmosphere, but water and carbon dioxide from volcanoes are considerably more abundant. The surplus outgassed water vapour has condensed into the oceans, but most of the CO_2 is dissolved in the oceans and then deposited as calcium carbonate in limestones. A small, but highly significant, part of the CO_2 is used in **photosynthesis**, whereby green plants use the energy of sunlight to convert H_2O and CO_2 into carbohydrates, whilst at the same time *releasing oxygen*. Most of the present atmospheric content of oxygen has been liberated by photosynthesis but, early in the Earth's history, another oxygen-releasing process, **photodissociation**, may have been important. Photodissociation involves the breakdown of water molecules by ultraviolet light from the Sun. Lightweight hydrogen molecules tend to escape the Earth's gravity field and be lost, but oxygen is retained, and a small proportion is converted into ozone which forms a shell in the upper atmosphere. This *ozone* reduces ultraviolet light by absorbing it, and so photodissociation is a self-regulating process because the oxygen produced ultimately prevents further dissociation. A few per cent of the present atmospheric content of oxygen may have been produced in this way (Shimizu 1976) but the rest required photosynthetic organisms.

Because oxygen was built up in the Earth's atmosphere from H_2O and CO_2 and, on this theory, was virtually absent from the *early* Earth, *reducing* conditions may have prevailed. Although the rate at which oxygen build-up occurred is much debated, the presence of relatively reducing conditions in Precambrian times is supported by evidence from sedimentary rocks (Fig. 10.7). The Archaean rock record contains banded ironstones composed of haematite (Fe_2O_3) and chert (SiO_2), thought to have been deposited by marine organic agencies. That iron reached marine environments demonstrates that it was liberated by weathering sites in the soluble Fe^{2+} state. In contrast, modern weathering produces insoluble Fe^{3+} which is incorporated into red, often aolian, sandstones, or **redbeds**. Thus, the change from banded ironstones to red ironstones about 2200 Ma ago must reflect increasingly oxidising surface conditions. The common presence of detrital minerals such as uraninite (U_3O_8) and pyrite (FeS_2) in Precambrian rocks, both of which are unstable in oxidising conditions, and the increasing abundance of sulphate deposits since about 2500 Ma ago, are also consistent with these ideas (see Towe 1978).

The earliest positive evidence for Precambrian life, in the form of microfossils, is recorded in some of the oldest sediments. But the earliest life-forms may even have preceded the formation of the oldest known rocks because the meta-sedimentary fragments found in the Amîtsoq gneisses contain ironstones that probably were   187 biogenically precipitated. If reducing conditions prevailed, then the lack of a strong ozone shield would have meant that life evolved in the presence of strong ultraviolet radiation, which is harmful because it destroys amino-acids. Possibly these early life-forms escaped the harmful effects by living in deep water. The simplest **procaryotic** cells (single cells or chains of cells, with no nucleus) derived their food and energy by fermentation. Later, they became photosynthetic and must have resembled modern blue-green algae. The discovery of quite complex procaryotic organisms in 3500 Ma rocks of the Swaziland System in South Africa by Knoll & Barghoorn (1977) indicates that photosynthesis may have started about this time, thus preceding by 1000 Ma the appearance of abundant oxidised sediments.

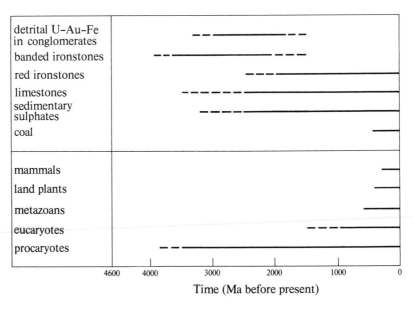

Figure 10.7 Distribution of fossil organisms and important environmental indicators in sedimentary rocks through time. (Dashed lines indicate uncertainties in the interpretation of the evidence.)

Opinions differ as to where life *originated*. The more traditional view, supported by successful experiments, is that a mixture of atmospheric gases was synthesised into large organic molecules by an energy source, such as lightning, and then evolved by poorly understood processes into primitive bacteria. A contrasting view is that life on Earth may even have had an extraterrestrial origin.

10.3   The Proterozoic

10.3.1   *Geology and tectonics*

According to Figure 10.2, most of the continental crust had formed by the end of Archaean times. Yet Proterozoic rocks, surrounding and covering the ancient Archaean cratonic nuclei, are considerably more abundant at the surface (Fig. 10.8). The shift in geological and tectonic patterns at the Archaean/Proterozoic boundary is quite marked; there are only rare occurrences of greenstone belts younger than 2500 Ma in age, and the early Proterozoic saw the development of extensive stable shield areas. Thick limestone–sandstone sequences of continental shelf type formed an extensive cover to the submerged Archaean basement, and the extent of these shallow Proterozoic ocean basins increased with time. The other dominant feature of the early Proterozoic crust, particularly in Canada and Australia, is the appearance of linear belts of trough sediments analogous to those of modern Cordilleran 'geosynclines' (back-arc basins or closing oceans, Section 9.3) where vast thicknesses of clastic quartzite–shale sequences accumulated. The oldest and best known, described by Hoffman (1973), is the Coronation geosyncline across northern Canada, where nearly 11 km of 2000 Ma old sediments accumulated, compared with 2 km on the adjacent craton in the same time interval. Similar linear zones of thick sedimentation are found in the African and Baltic shields where they are often intracratonic rather than marginal. The African mobile belts (for example, the Limpopo and the Ubendides in Fig. 10.8) may be the sites of former ocean closure, but the alternative view is that they represent 'ensialic' developments (i.e. founded on granite crust), with only restricted horizontal movements (Kröner 1979 gives a summary).

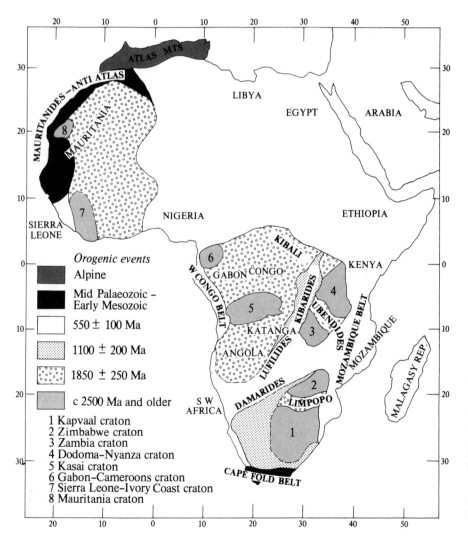

Figure 10.8 Major structural and time units in the geology of South and West Africa. Archaean cratonic nuclei are numbered and the ages of other crustal regions are indicated by shading. (Adapted from Clifford 1970.)

189

Evidence supporting the ensialic view is that some sedimentary sequences are easily traced from craton into mobile belt and that only rarely are calc-alkaline plate margin lavas found in the mobile belts. However, this may be due to their erosion, because most mobile belts are deeply eroded; calc-alkaline intrusives and lavas *are* found in low-grade belts with high levels of exposure. A particularly interesting example is the late Proterozoic Mozambique belt (Fig. 10.8) which has been interpreted as ensialic, where it is deeply eroded in the south, but carries andesite lavas, ophiolites and other evidence of subduction and suturing in the north, where it seemingly continues to form the crystalline basement of north-east Africa and Arabia.

In addition to their platform–geosyncline–mobile belt features, there are transcontinental basic dyke swarms, anorthosites, alkaline intrusive complexes and kimberlites (Fig. 7.1) within the Proterozoic continents. These are all related to magmagenesis deep in the mantle, and they indicate the relative stability of the crust, because the magmas found routes to the surface only where the crust was under tension. For example, the basic dyke swarms were developed in two major periods from 2500–2000 Ma and 1300–600 Ma ago: these ages correlate with the least active periods within mobile belts. Many of the massive anorthosites are associated with alkali-rich granites (adamellites) of 2000–1000 Ma age, which occur in high-grade granulite-facies metamorphic terrains (Fig. 10.9) suggesting that they were emplaced into previously metamorphosed and stabilised crust. Emslie (1978) suggested that these belts mark the sites of incipient continental rifting where tholeiitic magmas rose and penetrated the cratonic crust. Here they developed the anorthosite masses by flotation of fractionated plagioclase. The latent heat of crystallisation of the basic magmas was sufficient to cause partial melting at lower temperatures in crustal rocks. These new melts mixed with the residual tholeiitic magmas and formed the associated suite of alkali-rich diorite and adamellite intrusions. Similar alkaline complexes are known from younger intracontinental sites of rifting; for example, the 150 Ma old granites of Nigeria and the 50 Ma old granites of north-west Scotland. But the most prodigious period of alkali granite generation during the Earth's history was during the mid-Proterozoic (perhaps indicating that the continental crust resisted fragmentation at that time (Brown 1979)).

The evidence for plate margin developments around the cratonic nuclei becomes much more clear in later Proterozoic rocks. Examples are the Grenville belt in eastern Canada and the Pan-African belts of Arabia, Africa, South America, Australia and Antarctica (Fig. 10.9). The Pan-African 'episode', originally defined radiometrically, has now taken on a broader tectonic definition which encompasses all the Proterozoic rocks formed within and around the older African cratons in the period 1000–550 Ma. Black *et al.* (1979) found structural evidence for 800–600 Ma rifting and suturing along the eastern margin of the West African craton in Mali. Since the region was previously interpreted as an ensialic mobile belt, this evidence strengthens the case for modern-type plate tectonics in the late Proterozoic. In Libya, Egypt and Arabia, detailed studies have revealed calc-alkaline intrusive/extrusive lineaments separated by ultrabasic ophiolite zones, again indicating horizontal continental aggregation. The geochemical and isotopic characteristics of igneous rocks from the Arabian shield led Greenwood *et al.* (1976) and Gass (1977) to propose that the shield developed by magmatism over a series of adjacent island arcs much as those found today in the South-West Pacific. There is no indication that there was any continental crust in this region before 1000 Ma, and that the entire shield formed during the Pan-African period – a process now termed **cratonisation**.

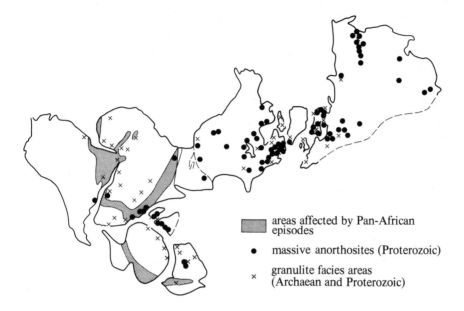

areas affected by Pan-African episodes

●　massive anorthosites (Proterozoic)

×　granulite facies areas (Archaean and Proterozoic)

Figure 10.9 Suggested late Proterozoic supercontinent based on palaeomagnetic data (Piper 1976) showing also the distribution of granulites, anorthosites and Pan-African belts. (Redrawn from Windley 1977. Used with the permission of John Wiley & Sons Ltd.)

How do the geological characteristics of the Proterozoic relate to the Earth's thermal state? The model of convection whereby the number of cells depends upon the heat production in the mantle, predicts fewer and larger cells as radioactivity declined during the Proterozoic. This implies that zones of magma generation would be more widely spaced and this may explain why large areas of continental lithosphere behaved in a stable fashion. The more surprising feature is the sheer size of these continental blocks compared with the Archaean high-grade domes. The mantle was convecting beneath the shields and there are signs that the lithosphere resisted fragmentation. At the margins of the Proterozoic continents, there were subduction zones associated with thick geosynclinal sedimentation, yet, within the continents, relative horizontal movements are uncertain until the late Proterozoic. There was no shortage of deformation and reworking within the continental regions, but the extent of lateral movements of continental lithosphere in the Proterozoic, as with the Archaean, remains an outstanding question.

10.3.2 Biological and atmospheric evolution in the Proterozoic

The evidence from sedimentary rock formations (Fig. 10.7) indicates a continued increase in atmospheric oxygen, probably due to more widely dispersed and abundant photosynthetic micro-organisms. Banded ironstones, such as the algal Gun Flint formation of Ontario, reached a peak of development on the borders of the Archaean cratons, between about 2600 and 1800 Ma ago. Red ironstones became more widespread, particularly in the late Proterozoic; for example, the thick red sandstone deposits of north-west Scotland and the Baltic shield. Early

191

Proterozoic conglomerates, containing detrital pyrite and uraninite, formed as Archaean greenstones were eroded and, to survive, these minerals must have escaped oxidation. But sedimentary sulphates and carbonates (limestones) were becoming more common because of progressive increases in the amounts of dissolved CO_2 and SO_2 (volcanic gases) in sea water. Like the banded ironstones, most of the early limestones were precipitated biochemically and they contain impressive numbers of procaryotic microfossils. Photosynthetic organisms must have been widespread in the Proterozoic and so oxygen was being released throughout. By about 1500 Ma ago, the atmosphere is thought to have contained about 1% of its present oxygen level, which is ample for the oxidation of sedimentary materials. A new kind of cellular micro-organism with a nucleus, like modern amoeba, developed during the Proterozoic. These are the **eucaryotes**, whose earliest positive identification is in the 1000 Ma old Bitter Springs formation of Australia (Fig. 10.7).

As the atmospheric oxygen content increased, so a more extensive ozone shield would have developed. A smaller depth of water would then be needed to protect organisms from ultraviolet radiation, so opening up a wider range of ecological environments in shallow water. Increased oxygen levels also allowed animals to develop which could breathe oxygen. The first traces of multicellular, **metazoan**, animal life are recorded in the late Proterozoic; for example, the 700 Ma old Ediacara fauna from sandstones of central Australia (Fig. 10.10). Among the various fossils are the soft-bodied impressions of primitive arthropods, sea-pens, jellyfish, worms and sea-urchins. Thus, the dawn of major biological diversity appears in the record of the late Proterozoic, for only within the last 600 Ma have most of the modern species evolved.

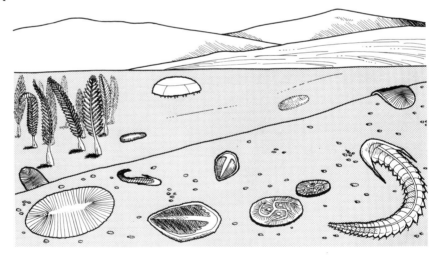

Figure 10.10 A reconstruction of the late Proterozoic Ediacara fauna from central Australia based on soft-bodied impressions in sandstones. (After A. Lee McAlester, *The history of life*, 2nd edn. © 1977, p. 24, Fig. 1–14. Reprinted by permission of Prentice-Hall Inc.)

10.4 The Phanerozoic

10.4.1 *Geology and tectonics*

There is little doubt that crustal processes in the last 570 Ma have been dominated by plate tectonics and so, to a large extent, Phanerozoic processes have been described in Chapters 7–9. However, only the youngest island 'Andean' and 'Alpine' arcs were described in Chapter 9; here, earlier examples of Phanerozoic arcs and sutures are considered which include the Urals of central Siberia and the Hercynian mountains of southern Europe. Both are thought to have developed about 250–300 Ma ago, but perhaps the best-studied Phanerozoic fold belt is the 400–500 Ma old Caledonian–Appalachian system whose imprint is found in most countries bordering the North Atlantic. The present Atlantic ocean is only some 100 Ma old and, before it opened, Europe and America were a united continent since Caledonian times. However, the evidence for a former Atlantic ocean of Lower Palaeozoic times, known as Iapetus, embraces faunal (Fig. 10.11), stratigraphic, geophysical and geochemical differences between the north-west and south-east forelands (details in papers by Phillips *et al.* 1976, Leggett *et al.* 1979, Simpson *et al.* 1979). Thick Lower Palaeozoic greywacke sandstone and shale sequences developed in the closing ocean basin; these sediments were deformed, upturned and thickened as the ocean closed, about 450 Ma ago, and formed accretionary sedimentary prisms on the two continental forelands in the manner of Figure 10.12. Ophiolite sequences were emplaced, both along the final suture through Newfoundland and Scotland (Fig. 10.11), and along zones where former island arcs of the Iapetus margins were accreted to the former continental forelands by closure of their marginal basins (e.g. in Anglesey, Wales). Cordilleran calc-alkaline intrusive and extrusive magmatism occurred on both flanks of these mountains and these have many similarities to, but some contrasts with, the modern magmatic arcs described in Section 9.3 (see Brown *et al.* 1980 for a more detailed analysis).

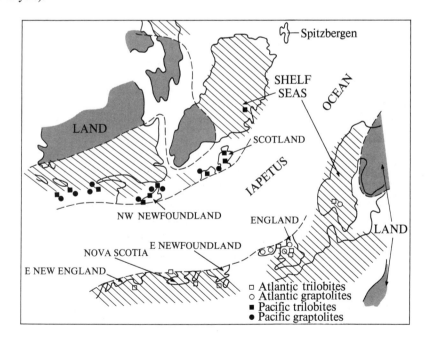

Figure 10.11 The Iapetus ocean in early Ordovician times with faunal evidence from Cowie (1974). The Iapetus ocean corresponded very roughly to the modern Atlantic in terms of its shorelines. (Redrawn from Windley 1977. Used with the permission of John Wiley & Sons Inc.)

Two important indicators of the force of plate collisions are first, the number of high-pressure, low-temperature **blueschists** (glaucophane schists, Fig. 9.5) which occur on leading continental margins and, secondly, the number of high-density ophiolite wedges trapped in back-arc closures and suture zones. There is good evidence that both became increasingly common in late Proterozoic and Phanerozoic times. This evidence, together with that of increasing deformation with time, suggests that continental plate movements have become *more* powerful during roughly the last 1000 Ma. So continental movements may have become more vigorous whilst mantle convection has become less vigorous – yet further evidence that the plates are moved *passively* by mantle convection (Section 8.8). Another feature of Phanerozoic tectonics is that about 200–300 Ma ago, after the Caledonides, Uralides and Hercynides had been formed by different continent–continent closures, all the continents were joined to form the supercontinent called Pangaea (Fig. 10.13) which, later, fragmented to give the modern continental configuration. The pattern of ocean openings and closings in the past has been far from simple, even in Phanerozoic times.

10.12 The development of an accretionary sedimentary prism of inclined sediments, coarsening upwards, at a subduction margin with motion of the subducted plate from left to right. (Adapted from Uyeda, S. 1978. *The new view of the Earth: moving continents and moving oceans.* © 1978, W. H. Freeman & Co.)

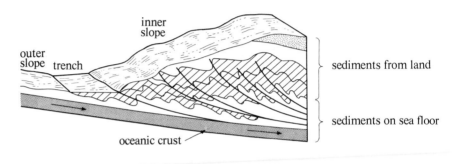

10.4.2 *Life in the Phanerozoic*

The preserved hard parts of animal fossils first occur in Cambrian rocks and this development marks the Proterozoic/Phanerozoic boundary. Thousands of new marine species appeared, and by 430 Ma ago land plants had evolved. Again, this may have been linked to the evolution of the atmosphere and the ozone shield, for some 10% of the present oxygen concentration is considered necessary for the survival of land-based species. By early Devonian times (380 Ma ago), dense plant-life had spread out over the land surface, and significant quantities of coal, produced by decaying vegetation, appear in the geological record (Fig. 10.7). The late Carboniferous saw the maximum development of massive forests and, by 270 Ma ago, atmospheric oxygen had probably reached its present concentration. So far as animals are concerned, amphibious creatures appeared in the Devonian as an evolutionary branch of the fish kingdom. As ecological niches on land diversified, amphibians gave rise to reptiles and eventually, by 250 Ma ago, to the mammals – although the latter did not become dominant until about 70 Ma ago.

The composition of the atmosphere has probably changed very little in the past 250 Ma and, today, there is a complex equilibrium in the atmosphere–hydrosphere–biosphere system dominated by (i) the production of oxygen from CO_2 by plants and (ii) the consumption of oxygen and regeneration of CO_2 by respiration of animals. In the last century, particularly, Man has also increased atmospheric CO_2 contents through the burning of fossil fuels (wood, coal, oil and natural gas).

10.5   Crustal mobility, growth and recycling

Earlier sections have shown that the tectonic pattern of the continental crust is discontinuous, varying from the small-scale Archaean pattern of granite–greenstone belts, through the platform–geosyncline–mobile belt pattern of the early Proterozoic, to the plate tectonic cycles which are well known today. Some of the unresolved issues concern: (i) the amount of relative continental movements at different times in the past, and (ii) the extent to which the continental crust has grown or been recycled throughout geological time. Additional evidence from radiogenic isotopes and palaeomagnetism bears on these problems.

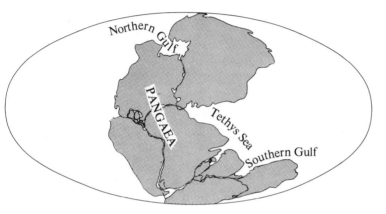

(a)

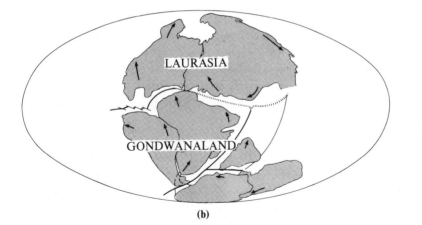

(b)

Figure 10.13 The distribution of modern continents (a) 200 Ma ago and (b) 180 Ma ago, just after fragmentation of the supercontinent, Pangaea, into Laurasia and Gondwanaland. Arrows in (b) indicate the direction of continental movements. (Adapted from Uyeda, S. 1978. *The new view of the Earth: moving continents and moving oceans.* © 1978, W. H. Freeman & Co.)

10.5.1   *Isotopes and crustal evolution*

Despite the difficulties of producing reliable Rb/Sr and Pb/Pb isochrons from ancient rocks (Notes 5 and 6), some remarkably good ages have been obtained, together with initial strontium isotope ratios. Figure 10.14 shows that the oldest Amîtsoq gneisses of western Greenland formed 3800 Ma ago, while the voluminous Nûk gneisses of the same area formed 2850 Ma ago. The Nûk gneisses were once interpreted by students of Greenland geology as having formed by the remobilis-

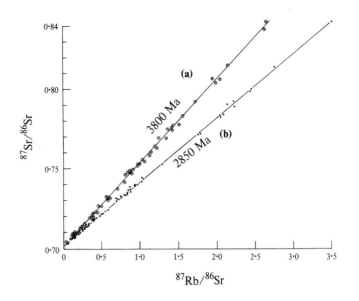

Figure 10.14 Rb/Sr whole
rock isochrons for (a) the
older Amîtsoq and (b) the
younger Nûk gneisses of
western Greenland. (After
Moorbath 1977. Used with the
permission of Elsevier
Scientific Publishing Co.)

ation of older, Amîtsoq-like Archaean crust but Figure 10.14 shows that both groups have low initial strontium isotope ratios. These initial ratios are plotted on an $^{87}Sr/^{86}Sr$ evolution diagram in Figure 10.15, together with extrapolated growth lines for the two types of material deduced from the *range* of their modern $^{87}Sr/^{86}Sr$ ratios. For the Nûk gneisses (point B) to have been formed from Amîtsoq precursors, their ages and initial ratios must plot within the triangle starting at the Amîtsoq origin (point A). This is not the case, and so the Amîtsoq and Nûk gneisses had *separate* origins. Using data for Archaean events in Greenland, Zimbabwe (Wilson *et al.* 1978) and North America, together with comparable U/Pb isotopic data, Moorbath (1977) concluded:

(a) that 'the low upper-mantle-type initial $^{87}Sr/^{86}Sr$ ratios of many ancient gneisses indicate that their immediate precursors are predominantly juvenile additions to the crust at, or very close to, the measured Rb/Sr whole-rock isochron age', implying that gneiss metamorphism postdates the formation of juvenile crust by only about 100 Ma,

(b) that mobile micro-continents with subduction zones (Fig. 10.5) provide the most suitable conditions for crust formation, and

(c) that the formation of later Archaean gneisses (e.g. the Nûk) represents another 'major, new addition to the continental crust', rather than being the product of reworking. In support of this interpretation, isotopic data from some K-rich granites, produced towards the end of activity in granite–greenstone regions (Fig. 10.5), have higher initial ratios, showing that crustal remelts can be distinguished from new additions to the crust (e.g. the Qôrqut granite of Greenland, point C, Fig. 10.15).

Ages have also been determined for other Archaean cratons and, like the Amîtsoq and Nûk gneiss ages, they fall into two quite distinct intervals, 3800–3500 Ma and 2900–2600 Ma before present. Using this information, various workers have argued

that major periods of crust formation are *isolated* in time; that is, they are **episodic**

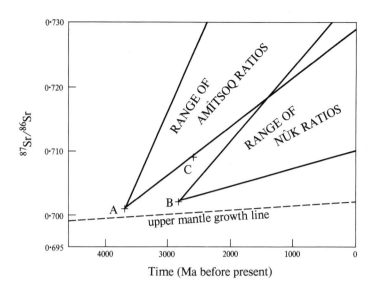

Figure 10.15 Strontium isotope evolution diagram for Amîtsoq (A) and Nûk (B) gneisses, showing the range of possible ratios generated by both, with time, based on their ranges of Rb/Sr ratios. Point C is the Qôrqut granite of western Greenland. (After Moorbath 1977. Used with the permission of Elsevier Scientific Publishing Co.)

rather than being continuous. Moorbath (1977) emphasised that this pattern of episodic growth may have continued throughout the Proterozoic and, so, to modern times, as indicated by the mantle-type initial ratios of magmas generated at young plate margins (Fig. 9.12, for example). For the Proterozoic, he cited isotopic evidence that the 'thermal events' seen in age groupings are the same for both continental margin and the intracontinental mobile belt zones. Most post-Archaean ages so far determined cluster into three age bands: 1900–1600 Ma, 1200–800 Ma and 600–0 Ma.

Referring now to the two models for the rate of crustal growth summarised in Figure 10.2 (model 1, crustal growth throughout the Earth's history; and model 2, steady-state crustal volume since the Archaean), a modified version of model 1, with a rather jerky curve, would fit these new data. Most continental growth would have occurred in Archaean times when heat productivity was high, leading to greater mantle melting and fractionation. This is consistent with the sea-water $^{87}Sr/^{86}Sr$ growth curve, determined from sedimentary rocks of all ages which shows an appreciable upturn about 2500 Ma ago (Fig. 10.16), indicating the growing influence of continental strontium as the volume of the continents and their erosion products increased (Section 10.2.1). Also in keeping with model 1, O'Nions *et al.* (1978, 1980) have argued that Sr, Pb and Nd isotopic data for mantle-derived basalts of all ages show progressive changes towards depletion of their source regions with time. This is best explained by **irreversible mantle differentiation**, either continuously, or in many stages, throughout the Earth's history as the continental crust increased in size.

There seems to be some agreement about *why* crustal growth should have been episodic rather than continuous. Even the proponents of model 2 (Fig. 10.2), who contest the interpretation of initial isotope ratios (see below), agree that age groupings represent periods of major crustal activity. Changes in tectonic patterns are attributed by most workers to changes in mantle convection patterns from smaller to larger cells as the Earth's internal heat sources are redistributed and refocused. These changes are thought to have caused renewed magmatism as

197

the number and wavelength of convection cells changed (Fyfe 1978).

It might seem that there is an almost irrefutable case for model 1 because mantle-type initial ratios are recognised from sites of crustal accretion throughout geological time. This is not so, for today a large number of geologists regard the volume of continental crust 2500 Ma ago as having been the same as (model 2a) or more than (model 2b) the present volume. They explain the observed isotopic ratios by processes such as (i) the subduction of vast quantities of erosion products back into the mantle, or (ii) isotopic exchange between crust and mantle magmas and/or remelting within low-Rb crust.

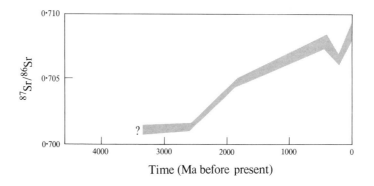

In the case of process (i), if sediments are subducted, they may carry ^{87}Sr-rich material into the zone of melting where mixing with juvenile mantle magmas takes place before eruption. This might provide an explanation for the trends towards increasingly 'mature' initial ratios observed, for example, in northern Chile (Fig. 9.12). However, mature strontium, neodymium and lead ratios, all reflecting a crustal component, occur in regions of thickened crust (Section 9.3.3) and do not necessarily correlate with the amount of continental detritus on adjacent oceanic crust (Hawkesworth *et al.* 1979, Brown 1979). Also, sediments are relatively light and may not be subducted deeply into dense mantle but may accumulate as accretionary prisms (Fig. 10.12) along continental margins. The proponents of sediment subduction believe that the density of the sediments is increased sufficiently by metamorphism in the initial stages of subduction (Fyfe 1981) and there are physical arguments (Molnar & Gray 1979) that, at least, small amounts of continental crust may be subducted. Thus, while continental detritus may be subducted, isotopic evidence suggests that this is not of major importance in terms of magmagenesis at the present day.

The second category of continental reworking processes involves crustal remelting from ^{87}Sr-depleted sources and/or isotopic exchange. In Sections 9.3.2 and 9.3.3, we concluded that crustal melts may be added to mantle-derived melts over modern subduction zones, but that, volumetrically, the latter are most important. In particular, it is difficult to see how the large volumes of Andean batholiths, or the Nûk gneisses, for example, could be obtained from relatively unfusible granulite-facies rocks in the lower crust. Finally, Fryer *et al.* (1979) have suggested that the initial ratios of continent-derived magmas may be reduced immediately after crystallisation by hydrothermal homogenisation with the strontium of mantle-derived rocks. In other words strontium migrates in hydrothermal fluids, resulting in isotopic exchange, until uniform ratios are achieved in the piece of crust affected. They consider this process to have been particularly important in the Archaean when, according to the thin lithosphere model (which they accept), most magmatic sites would have been submarine. But, if isotopic homogenisation within the crust is so readily achieved, then it is difficult to understand how the high initial strontium isotope ratios of post-tectonic intrusions (such as Qôrqut in Greenland, Fig. 10.15) could have survived. Also the *low* initial ratios of most young igneous rocks in modern zones of crustal thickening are not easily explained by isotopic exchange because an important source of mantle strontium is not available. Again, this implies the addition of *new* magmas to the crust. The process outlined in Section 9.3.3 involving a continuous supply of mantle-derived magmas, which, in particularly thick crustal zones, become progressively contaminated by crustal sources with time, seems most acceptable.

To conclude, it seems that crustal remelting, isotopic equilibration by hydrothermal activity and sediment subduction are all unable to explain fully the variations of magma initial ratios in zones of crustal accretion. The most reasonable explanation is that the crust has grown in volume progressively (model 1), perhaps episodically. If continental erosion products are resorbed into the mantle (model 2) they do not significantly affect magmagenesis. It seems likely that the sites of continental accretion are, and always have been, sites of continental growth.

10.5.2 *Palaeomagnetic polar wander curves*

The time-average direction of the Earth's magnetic field at the surface varies from vertical at the geomagnetic poles to horizontal at the magnetic equator (Fig. 6.2). On the assumptions that the field, on average, has been aligned with the Earth's rotation axis (i.e. the magnetic and geographic poles are, on average, coincident) and that the configuration of the field has remained constant, the palaeomagnetic techniques described in Section 6.2.1 can be used to determine the poles, thence the past positions of the continents from rocks whose ages are measured isotopically. The sequence of poles for a particular continent interval is known as a 'polar wander curve' since it was once thought that the magnetic poles 'wandered'. Now, of course, these curves are interpreted in terms of continental movements and, by examining polar wander curves for two continents, such as Europe and North America, in Figure 10.17, relative movements between them can be estimated. Both continents have similar curves for most of Phanerozoic times (Fig. 10.17a), which coincide for the Devonian to Cretaceous time interval (about 380–100 Ma, Fig. 10.17b) once they are brought into their best-fit positions. In the last 100 Ma, the opening of the Atlantic has caused the two curves to be separated at the young end. At the other end, the curves converge during the Cambrian to Devonian time interval (570–380 Ma ago) and this is a palaeomagnetic record of the closure of the Iapetus ocean (Fig. 10.11). The Cambrian poles for both continents lie near the equator, and to bring these back to the actual north pole we must move the continents south from their present positions. This implies that *both* North America and Europe have drifted north during Phanerozoic time and by amounts exceeding the relative movement between them. Palaeomagnetic data provide the basis for thinking that *all* the continents were united in the supercontinent, Pangaea, between about

199

200 Ma and 300 Ma ago (Fig. 10.13). This supercontinent broke up into northern (Laurasia) and southern (Gondwanaland) continents before fragmenting into the present-day continents.

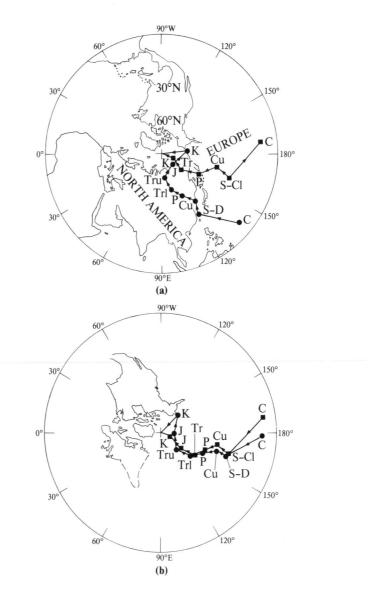

Figure 10.17 Polar wander curves for Europe (squares) and North America (circles) for Phanerozoic times with the continents in (a) their present-day positions and (b) their best-fit positions. Ages are: C = Cambrian, 550 Ma; S–D = Siluro-Devonian, 390 Ma; S–Cl = Silurian–Lower Carboniferous, *c.* 350 Ma; Cu = Upper Carboniferous, 300 Ma; P = Permian, 250 Ma; Trl = Lower Triassic, 220 Ma; Tru = Upper Triassic, 200 Ma; J = Jurassic, 150 Ma; K = Cretaceous, 100 Ma. (Redrawn from McElhinny 1973. Used with the permission of Cambridge University Press.)

The success of palaeomagnetism in defining continental movements during the Phanerozoic naturally stimulated attempts to do the same for the Precambrian. However, the technique is unable to resolve any movements between the Archaean high-grade domes. This is due to a combination of Proterozoic metamorphism (which caused some remagnetisation) and to the poor quality of age control. McElhinny and Embleton (1976) have shown that, if an isotopic age is known to an accuracy of only $\pm$ 100 Ma – as is often the case for early Precambrian rocks – and drift rates were 10 cm a^{-1}, a 5000 km ocean could form and close unnoticed within the statistical error of age determination. The entire age span of the polar wander curves in Figure 10.17, for example, is only 600 Ma. Much faster and smaller-scale relative movements are also to be *expected* according to the permobile model for Archaean crustal processes.

Although similar criticisms apply to Proterozoic palaeomagnetic studies, the larger scale of Proterozoic processes makes the technique potentially more useful. Piper *et al.* (1973) and Briden (1976) summarised the available data from the three major early Proterozoic cratonic areas in West Africa, the Congo and South Africa (Fig. 10.8). They concluded that the whole of Africa behaved as a *single*, but *mobile* (Fig. 10.18) plate during Proterozoic times and that significant latitudinal movement between the cratons may not have occurred. But they were unable to exclude the possibility of rapid intracratonic ocean openings and closings along the lines of mobile belts. Piper (1976) collated published Proterozoic palaeomagnetic data and recognised certain similarities in the shapes and sizes of the different polar wandering curves which suggest that *all* the continents might have moved as a united mass. He proposed that a Proterozoic supercontinent (Fig. 10.9) existed from 2250 to 1000 Ma ago, characterised by linear belts of massive anorthosites and granulites, associated with alkaline intrusive activity. This became particularly important as the supercontinent rifted and broke up in the late Proterozoic. These palaeomagnetic data are consistent with subduction-related continental accretion and cratonisation around the *margins* of the mobile supercontinent but are not consistent with major lateral movements across mobile belts.

A final important feature of polar wander curves arises from their shapes; smooth tracks with intermittent major excursions (Fig. 10.18), termed **kinks**, or **hairpins**. These represent major changes in the direction of continental movement, whereas the more linear parts of the polar tracks record periods of smooth, uniform changes in latitude (e.g. between 700 and 900 Ma in Fig. 10.18). There is a notable correlation between the kinks, or hairpins, and the timing of major rock-forming episodes according to isotopic data (Section 10.5.1). The two major periods of Proterozoic activity in Africa are represented by kinks between 1750 and 1950 Ma and again at about 1000 Ma in Figure 10.18.

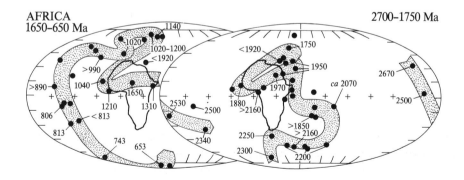

Figure 10.18 The Proterozoic polar wander path for Africa, indicating that the continent was mobile and that movement was complex. Data points represent various parts of Africa; they all fall on the same path, indicating that the continent behaved as a single plate during this time interval. (From Piper 1976. Used with the permission of the Royal Society.)

In answer to the points raised at the beginning of Section 10.5, we can say (i) There is good palaeomagnetic and isotopic evidence that wholesale changes in the direction of continental movement and magmatic activity were related in time. However, the cases for or against the permobile model for the evolution of the Archaean crust and rapid relative movements across intracontinental mobile belts

201

in the Proterozoic have not been proved. Palaeomagnetism indicates that any movements of the latter kind must have been quite rapid, and isotopic data favour the permobile micro-continent model for Archaean crustal evolution. The margins of the small Archaean continents provide ideal sites for the generation of juvenile magmas from the mantle. (ii) It is suggested that crustal growth by similar processes of subduction and melting have continued throughout the Earth's history. The magnitude of the process has diminished as the number of magma-penetration sites has decreased (Section 10.2.1), and the continental plates have grown larger. It is also suggested (cf. point 4 in the summary of Ch. 9) that growth of the subcontinental tectosphere root zones has kept pace with the volumetric increase of the continental crust. These tectosphere roots developed as partial melts were extracted from the mantle above subduction zones to form the continents.

10.6   A history of the Earth and a final summary

The final section of the book brings together the available evidence bearing on the evolution of the Earth as a whole; the outline that follows touches on many controversial points which have received detailed discussion earlier, and about which readers may have formed their own opinions.

About 4600 Ma ago, a supernova explosion took place in a nearby part of our Galaxy and added short-lived as well as stable isotopes to an interstellar cloud which consisted mainly of hydrogen (70%) and helium (28%), with the small balance comprising all the heavier elements. The pressure wave produced by the explosion may also have initiated the contraction of the cloud, which was probably large enough to form many stars.

A portion of the cloud contracted into the Solar Nebula, a rotating disc of matter. The bulk of its matter was at its centre, which contracted and heated up until it became hot enough to start hydrogen-burning and be a typical star: the Sun. The mixture of gas and dust in the outer parts of the Nebula was fractionated by the strong radial temperature gradient, so that material accreting to form the planets tended to contain progressively more low-temperature condensates further from the Sun. Most of the inner planets, including the Earth, formed in a region where light gases were depleted. A starting composition for the primitive Earth is provided by the least differentiated chondritic meteorites, which are believed to be fossils of the 'terrestrial' part of the Nebula, largely unaffected by later events.

But the Earth, being a more massive planetary body than the meteorite parent planets, became strongly heated in the late stages of accretion as the kinetic energy of the impacting and accreting planetisimals was converted to heat. The temperature was increased also by the decay of short-lived radioisotopes, such as ^{26}Al, inherited from the supernova. The first melt to form was a near-eutectic mixture of iron, nickel and sulphur, which collected – together with siderophilic trace elements – and permeated down through the more abundant 'mantle' silicates, because of its higher density, to form the core. The temperature of the separating mixture was probably above the eutectic because the composition of the core is more metallic than the eutectic. As the Earth cooled, very slowly, a nickel-rich Fe–Ni alloy, of high-temperature melting point, crystallised progressively to form the solid inner core (today, 1·7% of the Earth's mass) within the still-molten Fe–S outer core (about 30% of the Earth's mass).

Most of the core formed soon after the Earth accreted, when the heat of accretion and short-lived radioactivities were at a maximum, and probably was complete within 100 Ma. But the development of the crust, atmosphere, hydrosphere and biosphere took far longer and, in some cases, is still continuing at a significant rate. To start with, the mantle was not hot enough to melt, but its insulating properties are so good that the heat released by the long-lived radioactivities of its uranium, thorium and potassium – plus heat derived from the latent heat of inner-core formation – cannot be removed by conduction; consequently, the temperature rose until solid-state creep became sufficiently rapid to allow mantle convection. Mantle convection probably began fairly soon after the Earth formed and has tended to stabilise mantle temperatures ever since, but its vigour has probably declined with the exponential decrease of the radioactive isotopes, and the pattern of convection may have changed too.

Above the plastic convecting mantle, a rigid outer shell, or lithosphere, developed quite rapidly, with a base defined by the temperature at which mantle silicates deform inelastically, close to their melting temperature (about 1200°C for garnet lherzolite material which is undepleted). At the very beginning of the Earth's history, this lithosphere was so thin (a few kilometres) that it was totally unstable and it kept being resorbed into the mantle like the skin on molten metal. But by 3800 Ma ago, the first light and, therefore, unsinkable granitic crust had formed as the uppermost and most chemically fractionated constituent of the lithosphere. As the interior cooled slowly, the 1200°C isotherm moved deeper into the mantle and the continental lithosphere thickened. The actual thickness is rather indeterminate because of uncertainties about the distribution and focusing of heat flow. It is probable that rapid sea-floor spreading and subduction cycles disposed of the Earth's Archaean heat production, allowing small 'permobile' continents with thick lithosphere, perhaps about 100 km, to develop. On the other hand, the mantle may have convected beneath a lithosphere of uniform thickness which, to satisfy the higher Archaean heat production and heat flow, must have been thin (perhaps 10–30 km). The small nuclei of Archaean gneisses, separated by folded low-grade volcano–sedimentary greenstone sequences, contain high-pressure metamorphic rocks, indicating a thick crust and so favouring the permobile, thick lithosphere model (Fig. 10.5).

The atmosphere and oceans both appeared by 3800 Ma ago, due to prodigious outgassing, and the atmosphere contained large amounts of CO_2, N_2 and H_2O. The oceans provided a suitable environment in which early organisms developed about 3500 Ma ago, shielded from harmful ultraviolet solar radiation. There was little oxygen in the atmosphere at this time, but it was produced, first, by the photochemical dissociation of water and then by the photosynthetic activities of simple procaryotic organisms, such as blue-green algae.

Two major periods of Archaean crustal activity occurred at 3800–3500 and 2900–2600 Ma before present. A profound change in crustal tectonics, possibly due to changes in convection patterns in the cooling mantle, occurred about 2500 Ma ago, and the Archaean crustal fragments aggregated to form rigid plates which, by 2000 Ma ago, may have united into a supercontinent. Although the Proterozoic

203

continent(s) was mobile, platform-like and had marginal subduction zones and calc-alkaline magmatic arcs, it seems to have resisted fragmentation. Intracontinental anorthosites and alkaline intrusive complexes developed along possible rift zones related to transcontinental linear mobile belts. Sediments accumulated and were rapidly metamorphosed along those mobile belts which were *either* the sites of vertical swelling movements of the crust *or* the sites of short-lived lateral extensional, then compressional tectonics (i.e. small ocean openings and closings). Two further major periods of continental growth occurred in the Proterozoic at about 1800 Ma ago and again at about 1000 Ma ago. Around the continental margins, accretional processes became more common and, in particular, new, widespread Pan-African crust was established in regions were formerly only ocean crust had existed (this is the process of cratonisation).

Early in Phanerozoic times, there were several mobile continental plates whose behaviour was rather like modern plates, with marginal and intracontinental (collisional) magmatic arcs and orogenic belts. Another supercontinent, Pangaea, existed some 200–300 Ma ago, and this fragmented in Mesozoic times to give the present-day continental configurations.

The mantle is still evolving irreversibly, both chemically and physically, as new mantle-derived magmas contribute to the marginal growth of the continental crust in zones of oceanic lithosphere destruction, leaving depleted tectosphere roots in the mantle beneath. The continental crust is still growing, but at an ever-decreasing rate, though the extent to which erosion products are resorbed back into the mantle is unknown. Though radioactive heat production within the Earth must be declining, this does not necessarily mean that tectonic activity is, as yet, becoming more feeble. Some of the most intense thrust tectonics and high-pressure metamorphic rocks ever known have developed during the last 100 Ma – in the Alpine/Himalayan suture. The reasons for this are not yet fully understood: it might be because there have been changes in the nature of plate-driving forces (Section 8.8); for example, today it could be that the small number of plates and plate edges find relatively little resistance from the mantle at subduction zones.

Meanwhile, the increase of oxygen in the atmosphere during Proterozoic and early Phanerozoic times stimulated the evolution and increased diversity of life-forms. The first multicelled animals appeared in the late Precambrian while, in the Phanerozoic, life was able to leave the seas and exploit the varied evolutionary opportunities offered on land (see diagram inside rear cover). But this aspect of the Earth's evolution is not entirely dependent on crustal processes for its future development. One day, an unpredictably long time in the future, the Earth's interior will inevitably cool so much that convection, continental movements and hence mountain building, volcanoes and earthquakes, will all decline and cease. Then, erosion to base level may cause the surface to be covered with water. But will the evolution of the biosphere and atmosphere also came to a halt? Hardly likely, we suggest, though there will be radical changes by way of response to the new conditions. It is a sobering thought that human activities, which may themselves be causing irreversible changes to the environment, therefore, are mere transient geological phenomena.

General journal:

Schopf (1978): evolution of primitive organisms.

General books:

Windley (1977): crustal evolution (including an extensive bibliography).

Gass *et al.* (1971): various aspects of crustal evolution and the origin of life.

McAlester (1977); Laporte (1979): the origin and evolution of life.

Advanced journals:

Phil. Trans. R. Soc. Lond. **280A**, 1976, 397–667: 'A discussion on global tectonics in Proterozoic times', J. Sutton, R. M. Shackleton & J. C. Briden (eds).

Phil. Trans. R. Soc. Lond. **288A**, 1978, 383–646: 'Terrestrial heat and the generation of magmas', G. M. Brown, M. J. O'Hara & E. R. Oxburgh (eds).

Phil. Trans. R. Soc. Lond. (in press, 1981): 'The origin and evolution of the Earth's continental crust', S. Moorbath & B. F. Windley (eds).

Advanced books:

Condie (1976); Tarling (1978): various aspects of crustal evolution.

Postscript: the state of ignorance

In this book we have tried to describe what is known about the interior of the Earth and, in doing so, we have shown that there are many unresolved problems to which only speculative answers can be given at present. Therefore, this book could be regarded as a fairly precise summary of our state of ignorance!

Some of the problems which are outstanding and on which progress may be made in the next few years – an advance to a new state of ignorance? – are:

(a) more seismic information about the transitions in the upper mantle, the base of the mantle and the transition between the inner and outer cores;

(b) more precise estimates of core densities, particularly for the inner core;

(c) a better understanding of the process of magnetic field generation, including the distinction between energy sources due to radiogenic thermal convection and mechanical stirring by dense-particle crystallisation;

(d) more information on how the Solar System formed and the role that a supernova may have played in triggering it and contributing material;

(e) studies of the outer planets and their satellites to advance our knowledge of the Earth's accretion and its composition; and

(f) new geological, geophysical and geochemical understanding of:

    (i) the distribution of temperature inside the Earth.

    (ii) the mechanism of creep in the mantle,

    (iii) the form of convection in the mantle,

    (iv) the driving forces on tectonic plates,

    (v) high-pressure mineral stabilities in the crust and mantle, and

    (vi) the mechanisms of continental growth, past and present.

Notes

Note 1   Ray tracing and the seismic velocity–depth profiles

Equation (2.3) and Figure 2.4 indicate how a ray can be traced through the Earth, but we need some way of simplifying the method, to avoid carrying out a separate calculation at each of many interfaces along a ray path, for each of many rays. This is achieved by finding some parameter that does not change along the ray path and so is characteristic of a particular ray. Referring to Figure N1.1, we apply Equation (2.3) (Snell's law) at each interface, and obtain that

$$\frac{\sin i_{n-1}}{V_{n-1}} = \cdot \frac{\sin i_n}{V_n} \qquad\qquad \frac{\sin i'}{V_n} = \frac{\sin i_{n+1}}{V_{n+1}} \qquad\qquad \text{(N1.1)}$$

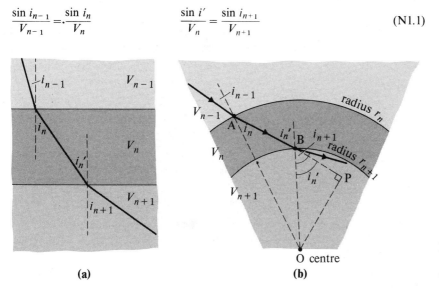

Figure N1.1   Refraction of rays through (a) plane and (b) concentric surfaces.

(a)

(b)

O centre

In Figure N1.1a, the interfaces are parallel and planar so that $i_n = i'_n$ and all the expressions are equal, i.e.

$$\frac{\sin i_n}{V_n} = \text{constant for all values of } n \qquad\qquad \text{(N1.2)}$$

But, in the case of the Earth, the layers are curved and i'_n no longer equals i_n (Fig. N1.1b), and therefore we have to find how they are related. From the figure, we can see that

$$\text{AO} \sin i_n = \text{OP} = \text{BO} \sin i'_n$$

i.e.

$$r_n \sin i_n = \text{OP} = r_{n+1} \sin i'_n \qquad\qquad \text{(N1.3)}$$

This is an additional condition applicable *within* each spherical layer which is considered to have constant seismic velocity. The transition *between* layers is still governed by Equation (N1.2), so the general condition is

$$\frac{r_n \sin i_n}{V_n} = \text{constant, } p \qquad\qquad \text{(N1.4)}$$

207

This equation is equivalent to both Equations (N1.2) and (N1.3), because at an interface, $r_n = r_{n+1}$ and Equation (N1.4) reduces to Equation (N1.2). Between interfaces, V_n is constant and it reduces to Equation (N1.3). Thus, it holds everywhere and the suffix n will be dropped.

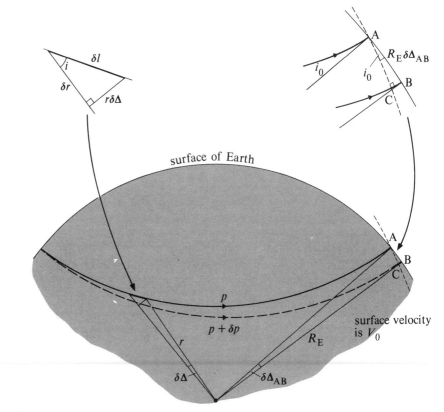

Figure N1.2 Ray tracing. Left and right sides illustrate detail of two areas of the main diagram.

The constant p has a simple physical significance. Figure N1.2 shows the paths of two rays which leave the source at nearly the same angle and so re-emerge at the surface close together. The ray that travels deeper will take longer, chiefly because it has to go a little further, by the length CB. This will take a time

$$\delta t = \mathrm{CB}/V_0 \tag{N1.5}$$

where V_0 is the seismic velocity just below the surface. From the geometry, we have

$$\sin i_0 = \mathrm{CB}/\mathrm{AB} \qquad \mathrm{AB} = R_E \, \delta_{\Delta_{AB}} \tag{N1.6}$$

and therefore, using Equations (N1.4), (N1.5) and (N1.6)

$$p = \frac{R_E \sin i_0}{V_0} = \frac{\delta t}{\delta_{\Delta_{AB}}} \rightarrow \frac{\mathrm{d}t}{\mathrm{d}_{\Delta_{AB}}} \tag{N1.7}$$

As $\mathrm{d}t/\mathrm{d}\Delta$ is simply the slope of the travel-time curve (Fig. 2.4c), the value p of a ray which emerges at some epicentral angle Δ is equal to the slope of the travel-time curve at this angle. Therefore p is the reciprocal of the *apparent* velocity, i.e. the velocity at which the seismic arrival appears to move along the surface of the Earth, as measured by the time of arrival at successive stations situated on a great circle passing through the origin of the circle. This fact will be used later.

Since p is the same at any point along a ray, it can also be expressed for the deepest point, r_d, where the ray is travelling horizontally, i.e. $i = 90°$, giving

$$p = \frac{r_d \sin 90°}{V_d} = \frac{r_d}{V_d} \qquad \text{(N1.8)}$$

Since p is known from Equation (N1.7), we can evaluate r_d/V_d. But to solve for either, a further equation for either r_d or V_d is needed, and this we get by tracing a ray along its path.

Referring to Figure N1.2, for any short segment δl of a ray path which subtends an angle $\delta\Delta$ at the centre of the Earth, we can write

$$\delta l^2 = \delta r^2 + (r\,\delta\Delta)^2 \qquad \text{(N1.9)}$$

and, from Equation (N1.4)

$$p = \frac{r \sin i}{V} = \frac{r}{V} \frac{r\delta\Delta}{\delta l} \qquad \text{(N1.10)}$$

Combining these two equations gives

$$\delta\Delta = \frac{p\,\delta r}{r[(r/V)^2 - p^2]^{\frac{1}{2}}} \qquad \text{(N1.11)}$$

This tells us how far a ray – characterised by its value of p – travels tangentially ($\delta\Delta$) as it descends a distance δr. If all the $\delta\Delta$ values are added, it will give the total epicentral angle travelled and this must be equal to a corresponding addition of the right-hand side for each increment of the path. In the calculus limit, this is

$$\Delta = 2 \int_{r_d}^{R_E} \frac{p}{r[(r/V)^2 - p^2]^{\frac{1}{2}}}\, dr \qquad \text{(N1.12)}$$

The ray starts at R_E, descends to its deepest point r_d and then returns symmetrically to the surface, hence the factor of two.

This is a mathematical expression for ray tracing, for it would allow the position of emergence at the surface, Δ, of a given ray, p, to be calculated provided the velocity were known at all depths.

This is roughly the inverse of what we want and the expression has to be inverted to a more useful form. This is done by mathematical manipulation, the details of which are given in Richter (1958, Appendix VI) or, briefly, in Stacey (1977, Appendix D). The result is

$$\log_e \frac{R}{r_{d_1}} = \frac{1}{\pi} \int_0^{\Delta_1} \cosh\left[\frac{p}{(r/V)}\right] d\Delta \qquad \text{(N1.13)}$$

The quantity in square brackets refers to a particular ray characterised by some particular value of p and can be evaluated using Equations (N1.7) and (N1.8). It is then a simple, if time-consuming, matter to draw a graph of $\cos h\,[p/(r/V)]$ against Δ, the position of re-emergence of the ray. If the area under this curve is integrated up to some particular epicentral angle Δ, every quantity in Equation (N1.13) is known except r_{d1}, which accordingly is found from the equation. This is the maximum depth of the ray which emerges at Δ_1, and Equation (N1.8) then allows us to calculate the corresponding velocity. By integrating up to progressively larger values of Δ_1, we can deduce the velocity at progressively greater depths; and so construct a velocity–depth profile. This can be done for both P- and S-waves.

Note 2   Moment of inertia, angular momentum, etc.

If Newton's laws of motion are extended from linear motion to rotary motion, it is natural to replace distance moved by the angle through which a body rotates, velocity by angular velocity, force by torque, and so on. The relation

$$\text{force} = \text{mass} \times \text{acceleration}$$

can then be replaced by a corresponding one

$$\text{torque} = I \times \text{angular acceleration}$$

where I is a new quantity chosen to make the relation hold, and is called the moment of inertia (Fig. N2.1).

Figure N2.1 Moment of inertia of body rotating about any axis.

The moment of inertia of a small (point) mass δm about an axis at a distance r away turns out to be $r^2 \, \delta m$, and so for an extended body it is

$$I = \quad \Sigma r^2 \, \delta m \ = \ \int r^2 \, \rho \, \mathrm{d}v$$

$$\text{all } \delta m \qquad \text{vol}$$

Note that I depends upon the position of the axis as well as upon the size and density of the body. I is analogous to mass (see the equation above).

Table N2.1 Linear and rotary motion

Linear quantities		*Angular quantities*	
linear displacement	x	θ	angular displacement
linear velocity	$V = \mathrm{d}x/\mathrm{d}t$	$\omega = \mathrm{d}\theta/\mathrm{d}t$	angular velocity
linear acceleration	$a = \mathrm{d}V/\mathrm{d}t$	$\dot{\omega} = \mathrm{d}\omega/\mathrm{d}t$	angular acceleration
force	F	Γ	couple or torque
mass	M	$I = \Sigma mr^2$	moment of inertia
momentum	mV	$I\omega$	angular momentum
	$F = ma = \dfrac{\mathrm{d}}{\mathrm{d}t}(mV)$	$\Gamma = I\dot{\omega} = \dfrac{\mathrm{d}}{\mathrm{d}t}(I\omega)$	

The last line of the table shows that momentum, linear or angular, remains constant or is conserved if no *external* forces act upon a body. Thus, if a body contracts due to internal forces, such as its own gravitational attraction, angular momentum is conserved, and since I must get smaller, ω must increase. This fact is used by a skater spinning on her toe: if she flings out her arms she spins more slowly; when she brings them close to her body she speeds up again. The conservation of angular momentum is important when considering the formation of the Solar System in Chapter 4. It shows that a contracting nebula of gas must rotate more rapidly as it grows smaller.

Note 3   Mean atomic weight ($\bar{m}$)

This parameter was introduced by Birch in 1961. From Chapter 3, seismic velocities provide only two equations involving the three variables K, μ and ρ, and, to get a unique solution, a third relation is needed. The attempt by Adams and Williamson to do this by assuming that density depends only on self-compression failed when it was found to be untrue for parts of the Earth. A different approach is to find a direct relationship between seismic velocity and composition.

Birch (1961) measured the seismic P-wave velocity of a large number of minerals and found the empirical relation:

$$V_p = a\bar{m} + b\rho$$

where $\bar{m}$ and ρ are the mean atomic weight and density, respectively, and a and b are constants.

This relation shows – to the extent that the equation holds – that the seismic velocity does not depend on the chemical bonds between the atoms that make up a mineral but only upon their mean weight. Nor does the detailed structure of the mineral lattice matter except insofar as it affects the separation of the atoms, which in turn affects the density. Thus, changes in volume caused by pressure, temperature or phase changes are all taken into account by the density term, while composition differences affect both terms.

Though it has been shown that Birch's law holds quite well for nearly all minerals examined, the underlying physical reasons for this have not been well understood. A number of persons have suggested that, for example, at very high pressures a power law would be a better description. In particular, Chung (1972) and Davies (1977c) argued that the linear approximation would begin to break down at the higher densities produced by very high pressures. This has been found, for instance, by Liebermann and Ringwood (1973), who have shown that the law is not fully obeyed during polymorphic transitions like those discussed in Section 7.4. Anderson (1973) has suggested a form of the power law that encompasses these high-pressure transitions.

The use of Birch's law has been largely superseded by the results of experimental petrology, but it is still useful for the lower mantle, where data are scarce. The value of the parameter $\bar{m}$ is that, being relatively insensitive to chemical composition, it provides a generalised description of the composition; conversely, it follows that even an exact knowledge of $\bar{m}$ would not yield details of what minerals or elements are present. Further details are given in Section 7.5.

Note 4   Magnetic braking

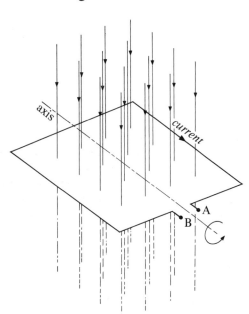

Figure N4.1 Magnetic flux threading a loop of wire.

Consider a loop of wire with a small gap, AB, placed in a magnetic field (Fig. N4.1). There is a magnetic flux through the loop which is simply the number of lines of magnetic force, or induction, which thread through the loop. Suppose that the flux changes, either because the

211

field changes or because the loop is moved, then an electromotive force (e.m.f.), or voltage, is induced between A and B, proportional to the rate of change of flux (Faraday's law):

$$\varepsilon = \frac{-\mathrm{d}\phi}{\mathrm{d}t}$$

If A and B are joined, a current flows proportional to the e.m.f., and this is the basis of the generation of electricity in dynamos and generators.

The electrical power cannot be had for nothing, and work has to be done in rotating the loop. This arises because there are forces upon a wire carrying a current when it is in a magnetic field, and these forces oppose the motion. An alternative way of regarding it is that the current in the loop produces a field like that of a magnet (Fig. N4.2) and will try to twist until it is aligned along the lines of force: this twist is opposite to the rotation of the loop.

Note that the magnetic field, and hence flux through the loop, *produced by the current in the loop* is greatest when the external flux is *changing* most rapidly, and is in such a sense that the total flux change through the loop is less than would be the case if the current did not flow. If the loop has no electrical resistance, this compensation will be perfect. (This can be deduced because Ohm's law tells us that the e.m.f. is equal to the product of the resistance and the current: if the resistance is zero, then so must be the e.m.f. But Faraday's law, above, states that the flux change is proportional to the e.m.f., so it too must be zero.) Therefore, in a perfect conductor, a current is induced whose associated flux through the loop exactly equals the change of external flux, so that the total flux remains constant.

If the simple loop were replaced by an arrangement of interconnected conducting loops of any shape, this effect would ensure that the flux threading every one of them would be unchanging. This can be generalised to a continuous, perfect conductor, such as a block; in it there can be no flux change, appropriate currents being induced to ensure this.

Imagine that we try to rotate part of a perfect conductor, say a cylinder, within the rest. Currents are induced to keep constant the flux within the cylinder, and forces are produced opposing the rotation, just as with the loop (Fig. N4.2). Thus, in general, the presence of a magnetic field in a conductor tends to prevent relative movements within it, and therefore effectively stiffens it.

These conclusions are applicable to an interstellar cloud or other gas containing ions, for such an ionised gas is a very good conductor. This may seen surprising for such a tenuous material, but it arises because the ions can move easily, with few collisions; as a result, even a very small electric field, or voltage difference, will cause them to accelerate and reach high velocities, and the flow of charges forms a current. To the extent that some collisions do occur, the cloud is not a perfect conductor and so flux changes can occur, albeit slowly. The conclusion remains that a magnetic field in a fluid conductor stiffens it, and opposes relative motions within it.

Figure N4.2 This uniform flux induces a current in the loop; in turn, this current produces a magnetic field – shown by the curved lines – which is similar to that of a bar magnet.

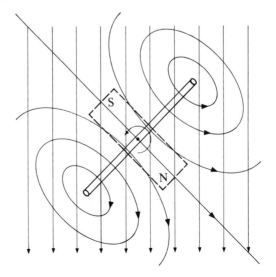

Note 5   Rubidium–strontium dating and strontium isotope initial ratios

^{87}Rb is radioactive and decays to ^{87}Sr; ^{86}Sr also exists but is not formed by decay and so can be used as a reference. Consider a magma that is completely homogeneous isotopically and which then forms into separate minerals. Different minerals in a given rock will have different Rb/Sr ratios, though all will have the same ^{87}Sr/^{86}Sr when they form. For example, biotite and K-feldspar have much higher Rb/Sr ratios than pyroxene and Ca/Na-feldspar. In Figure N5.1, four minerals are shown with compositions A_0, B_0, C_0 and D_0 at the time the rock crystallised. Now suppose the minerals remain closed subsystems in which the only change is the increase of ^{87}Sr due to the decay of ^{87}Rb, so that A_0 evolves to A_1, B_0 to B_1, and so on. The increase in the ^{87}Sr/^{86}Sr ratio will be proportional to the Rb/Sr ratio of each mineral (A–D), obeying the equation

$$\left(\frac{^{87}\text{Sr}}{^{86}\text{Sr}}\right)_{\text{now}} = \left(\frac{^{87}\text{Sr}}{^{86}\text{Sr}}\right)_{\text{initial}} + \left(\frac{^{87}\text{Rb}}{^{86}\text{Sr}}\right)_{\text{now}} (e^{\lambda t}-1)$$

where λ is the decay constant of ^{87}Rb ($1\cdot42 \times 10^{-11}$ a^{-1}). (For all geological uses of Rb/Sr dating, the expression $e^{\lambda t} - 1$ may be simplified to λt.) At a given instant, λt will be the same for all four minerals (t being the time elapsed since formation) as, of course, is $(^{87}$Sr/^{86}Sr$)_{\text{initial}}$. Thus A_1, B_1, C_1 and D_1 will all lie on a straight line or **isochron** (literally 'equal time') of the form

$$y = (\text{constant}) + x \,(\text{slope})$$

The age of the rock (t) may be calculated from the slope of the isochron which is, simply, λt. Note that the isochron method allows the age to be deduced despite the presence of an unknown amount of strontium with an unknown isotopic composition when the system closed.

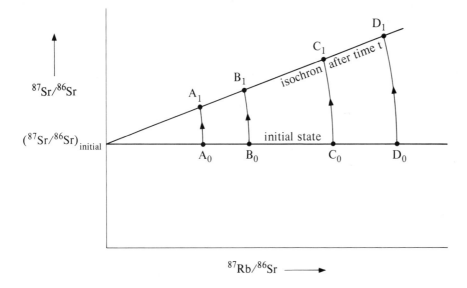

Figure N5.1 An example of an Rb/Sr isochron plot for a rock containing four minerals whose present-day ratios define the line (isochron) A_1–D_1. All minerals had the same *initial* ^{87}Sr/^{86}Sr ratios (A_0–D_0) when the rock formed. The initial ratio can be found from the intercept of the isochron on the y-axis and the age of the rock is found from the slope (λt) of the isochron.

The *initial isotopic composition of strontium* is determined from the intercept of the isochron on the y-axis and provides information about the source region of the magma being dated. It is believed that the ^{87}Sr/^{86}Sr ratio of the whole Earth, when first it formed, was the same as that of basaltic achondrite meteorites: $0\cdot699$. The ratio in modern oceanic basalts (from the mantle) is about $0\cdot703$–$0\cdot704$, so that the ratio has grown from $0\cdot699$ to about $0\cdot704$ during the age of the Earth. This would be expected from the average Rb/Sr ratio of mantle rocks which is about $0\cdot02$. Its change with time, or **growth curve**, is line AB (Fig. N5.2).

(Minor deviations from the present-day ratio of 0·704 occur due to mantle inhomogeneities, see Section 9.3.3.)

Now consider crustal material extracted from the mantle long ago. Immediately after extraction, its $^{87}Sr/^{86}Sr$ was the same as the mantle because isotopic ratios are hardly affected by the chemical processes of extraction. But the Rb/Sr ratio of this crust is much greater than that of the mantle, and it will follow a much steeper growth curve (Fig. N5.2). The origin of the particular crustal growth curve depends on the time of extraction and its slope is determined by the Rb/Sr ratio. The two curves shown are for crust stabilised 3000 Ma ago and 500 Ma ago: both have the same Rb/Sr ratio of 0·3. Clearly, crustal growth curves rapidly exceed the present-day mantle ratio and this permits us to distinguish between a mantle and a crustal magma source.

Consider, for example, the melting of mantle or crustal material with present-day ratios. Magmas with simple *mantle derivation* cannot have initial $^{87}Sr/^{86}Sr$ ratios greater than 0·704, but *crustal melts* will reflect the higher ratios reached by their older parental crust, often considerably in excess of 0·71. A wide variety of possible crustal melt ratios exists and, often, mixed crust and mantle sources complicate the issue. But rocks with initial ratios less than or equal to 0·704 must have been derived from the mantle.

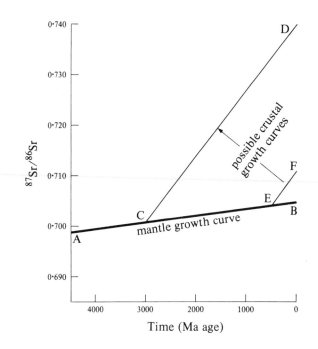

Figure N5.2 Schematic strontium isotope growth curves for the Earth's mantle (AB) and crustal rocks of two ages: 3000 Ma (CD) and 500 Ma (EF) old. The initial ratio for a rock generated by mantle melting today will be about 0·704, whereas a rock generated by melting 3000 Ma old crust would have an initial ratio of about 0·740. (Crustal growth curves are based on Rb/Sr = 0·3 – in nature, this ratio varies from about 0·05 to 1·0.)

Note 6   The lead/lead dating methods

The lead/lead (Pb/Pb) method is analogous to the Rb/Sr one (Note 5), but is more complex for it depends upon the decay of *two* isotopes of the parent to *two* isotopes of lead:

$$^{238}U \;\; \rightarrow \;^{206}Pb \quad \text{half life } 4{\cdot}5 \times 10^9 \text{ years}$$
$$^{235}U \;\; \rightarrow \;^{207}Pb \quad \text{half life } 0{\cdot}70 \times 10^9 \text{ years}$$

A third isotope of lead, ^{204}Pb, is not formed by decay and so (like ^{86}Sr) can be used as a reference isotope.

Consider a system which initially is isotopically homogeneous, and which then divides into subsystems, each with a different U/Pb ratio. The plot employed is different from that used for the Rb/Sr case, and consists only of ratios of lead isotopes, and so all subsystems plot at the same point before any uranium decay has occurred (Fig. N6.1).

Note 6   The lead/lead dating method

Decay of uranium increases both ratios, and more rapidly the greater the U/Pb ratio. Because the two isotopes of uranium decay at different rates, evolution follows a curve but, at any subsequent instant, the compositions of different subsystems lie on a straight line (isochron), whose slope (as plotted in the figure) decreases with time.

Lead from several different stony meteorites, from the troilite (FeS) phase of iron meteorites, from lunar and from terrestrial rocks all lie close to a straight line with a slope corresponding to about 4.6×10^9 years. This shows that these bodies separated form a common system at that time. Though often referred to as the 'age of the Earth', it was really when the various bodies of the Solar System became separate entities, though probably not yet in their present forms.

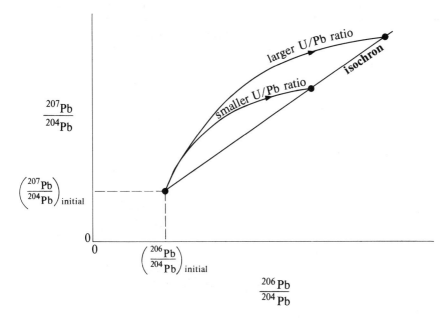

Figure N6.1 Lead isotope growth curves and an isochron.

Note 7   The atom and the nucleus

The chemical properties of an **element** depend upon the number of electrons surrounding the nucleus. This number equals the number of protons in the nucleus; **isotopes** differ in the number of neutrons in the nucleus, which, however, roughly equals the number of protons.

The properties of the **nucleus** depend upon the total number of **nucleons** (protons and neutrons). Nucleons are held together by strong nuclear forces which, however, operate only between adjacent nucleons. Consider the number of bonds in a nucleus: with two nucleons there is one bond (Fig. N7.1a); with three, three bonds; with four, six bonds. Thus the number of bonds *per nucleon* increases rapidly. But this increase does not continue because, once the nucleus reaches a certain size, nucleons are no longer adjacent to all the other nucleons and then each additional nucleon adds the same number of bonds (Fig. N7.1b). Since each bond made releases energy, this explains why the binding energy per nucleon curve (Fig. 4.12, curve a) at first rises steeply, then less so.

In addition, *all* the protons repel each other, because electrical repulsion falls off only as the inverse square of the distance. This contributes a *negative* binding energy per nucleon which continues to increase as the nucleus grows. This accounts for the decrease in total binding energy per nucleon beyond iron. (Other forces, which will not be discussed, account for the details of the curve.)

The nucleus becomes unstable long before the total binding energy reaches zero, for reasons peculiar to atomic particles (wave mechanics). A heavy nucleus may adjust to greater stability by emitting an α-particle or, alternatively, by splitting roughly into two (fission). Fission is important only for nuclei heavier than uranium, though it can be induced in ^{235}U and is the basis of nuclear power. Fission of ^{244}Pu yields xenon isotopes, amongst other fragments, and was met in connection with meteorite formation ages (Section 4.5.4).

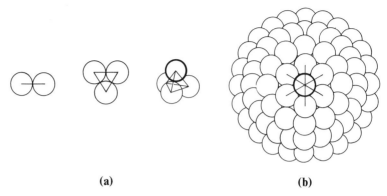

Figure N7.1 Nuclear bonds.   **(a)**   **(b)**

Note 8   Resistive heating and the efficiency of the geomagnetic dynamo

The generation of the Earth's magnetic field by a dynamo depends upon mechanical motions in the outer core powered either by thermal or gravitational convection. In a conventional dynamo, any resistive (ohmic) heating due either to current flowing through the windings or to currents induced by the changing magnetic field represents an energy loss which lowers the efficiency of the conversion from mechanical to magnetic energy. However, this may not be the case for the geomagnetic dynamo.

Because the regions of the Earth outside the core have very low electrical conductivities, only very small currents can be induced there and hence the energy loss must be small. Consequently, all ohmic dissipation must take place within the core itself, where it will cause heating. Depending upon the extent of heat losses to the mantle and the configuration of the dynamo, this heating may enhance the convective motions, so tending towards perfect efficiency, or perpetual motion. Near-perfect energy conversion also applies to the component of gravitational convection arising from the partitioning of iron–nickel alloy from the outer to the inner core, which is equivalent to the gravitational settling of a denser component.

Note 9   Heat flow

The rate at which heat flows out of the Earth could, in principle, be measured by standing a saucepan of water on the ground and observing the rate at which it warms up. However, the rate is so slow that it would take about two months for the temperature to rise 1°C – supposing the saucepan was unaffected by variations in the temperature of its surroundings, and so on. Clearly, this is not a feasible method, and heat flow, dQ/dt, is determined *indirectly*, by combining a knowledge of the vertical temperature gradient, dT/dr, and the thermal conductivity, k, of the rock:

$$\frac{dQ}{dt} = k\,\frac{dT}{dr}$$

The temperature gradient is measured using sensitive thermometers (usually thermistors) at known spacing. On land, this has to be done in boreholes, and precautions have to be taken to avoid water flowing through the hole and the disturbing effects of drilling. Generally, a concrete-lined hole at least 500 m deep is needed, and, as these are expensive, there are few land-based heat flow measurements. Conductivities are determined either in the laboratory using samples from the borehole, or determined *in situ* by measuring the rate at which the temperature rises when a heater is placed in the hole.

On the ocean floor, there is usually sufficient thickness of unconsolidated sediments to allow a hollow probe several metres long to be pushed in. This carries thermistors along its length to measure the temperature and hence its gradient, while the body of the probe acts as a corer and removes a sample whose conductivity is measured at the surface. Because this is a much easier process than using boreholes, there are many more heat flow measurements for the oceans than for the continents, though the results may be less reliable because of water circulating in the soft sediments, etc.

Further details can be found in standard textbooks such as Garland (1971), Jacobs *et al.* (1974), and Press & Siever (1978).

Note 10   Earthquake fault-plane solutions

When a region containing a fault is subjected to a shear stress – perhaps by the relative movement of plates – strain builds up in the surrounding area until the fault can no longer resist movement. It yields abruptly, releasing the stored strain energy as an earthquake, and generates seismic waves. Figure N10.1, which fits observation of the 1906 San Francisco and other earthquakes, depicts the sequence of events.

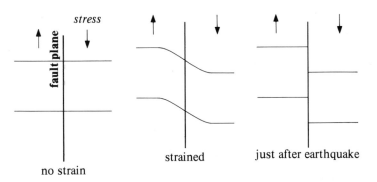

Figure N10.1 Deformation associated with an earthquake.

Any sudden movement of the ground generates seismic waves, but it is important to realise that the motion they produce at a distant point depends partly on the position of that point (Fig. N10.2). The movement of S causes compression in the direction of A, so that the *first* effect at A is the passage of a **compression**, followed by a series of rarefactions and compressions as the rest of the wave train passes by (cf. Fig. 2.1a); at B the *first arrival* is a rarefaction or **dilatation**. Thus A and B both receive P-waves, though of opposite sign. Since there is no movement in the direction of C, it does not receive P-waves, but it does receive an S-wave whose first motion is parallel to the movement of S. In the general case, D, both P- and S-waves are received.

Figure N10.2 Motions produced about a displaced point S.

Now consider the effect of movement on the fault, shown by the half-arrows (Fig. N10.3). As far as P-waves are concerned, L receives none, since seismic movement at the focus, F, is transverse to the direction FA. M does not receive P-waves either, because, though each side of the fault generates motion along FM, the resulting waves are of opposite sign and cancel at M. At other places, such as N, the effect of the far side of the fault is less than that of the near side because of the intervening fault plane, and it receives a P-wave (and also an S-wave). Similar arguments apply in other directions, and alternate quadrants receive compressive ($+$) and dilational ($-$) first arrivals. The length of the arrows within the lobes indicates the relative *magnitude* of the first arrivals in the direction of the arrows. (S-waves have a similar distribution, though rotated through 45° with respect to that for the P-wave.)

By observing the sense of first movements at seismic stations distributed around the world, it is possible – after making allowances for curvature of ray paths in the Earth – to deduce the pattern. Note that there is an inherent ambiguity: the fault plane could be as shown with dextral movement (as observed of one side of the fault from the other), but the same pattern would result from a fault plane FL with sinistral movement. In practice, it is usually easy to decide between these and so deduce the sense of movement on the fault and of the stress producing it.

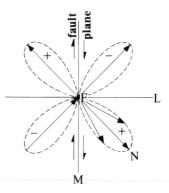

Figure N10.3 Magnitude and sign of first motions (P-waves) due to an earthquake at F.

Thus, by observing the first motions of the ground, due to the passage of P- and S-waves at a number of seismic stations (Fig. 2.2), it is possible to deduce not only the location of an earthquake but also the direction of the associated fault plane and the sense of movement across it. In this way it is found, for instance, that many fault planes are not vertical.

Finally, from this knowledge it is possible to estimate the directions of the stresses causing the earthquake.

218

References

Adams, F. D. 1938. *The birth and development of the geological sciences.* London: Baillière, Tindall & Cox.

Adams, R. D. and M. J. Randall 1964. The fine structure of the Earth's core. *Bull. Seism. Soc. Am.* **54**, 1299–313.

Ahrens, T. J. 1979. Equations of state of iron sulphide and constraints on the sulphur content of the Earth. *J. geophys. res.* **84**, 985–98.

Alfvén, H. 1954. *On the origin of the Solar System.* Oxford University Press.

Alfvén, H. 1978. Origin of the Solar System. In *The origin of the Solar System*, S. F. Dermott (ed.), 19–40. Chichester: Wiley.

Alterman, Z., H. Jarosch and C. L. Pekeris 1959. Oscillations of the Earth. *Proc. R. Soc.* **252A**, 80–95.

Anderson, D. L. 1973. Comments on the power law representation of Birch's law. *J. geophys. res.* **78**, 4901–14.

Anderson, D. L. and R. S. Hart 1976. An Earth model based on free oscillations and body waves. *J. geophys. res.* **81**, 1461–75.

Armstrong, R. L., W. H. Taubeneck and P. O. Hales 1977. Rb–Sr and K–Ar geochronometry of Mesozoic granitic rocks and their Sr isotopic composition, Oregon, Washington and Idaho. *Geol. Soc. Am. bull.* **88**, 397–411.

Bateman, P. C. and F. C. W. Dodge 1970. Variations of major chemical constituents across the central Sierra Nevada batholith. *Geol. Soc. Am. bull.* **81**, 409–20.

Bateman, P. C. and J. P. Eaton 1967. Sierra Nevada batholith. *Science* **158**, 1407–17.

Benioff, H. 1958. Long waves observed in the Kamchatka earthquake of November 4, 1952. *J. geophys. res.* **63**, 589–93.

Bickle, M. J. 1978. Heat losses from the Earth: a constraint on Archaean tectonics from the relation between geothermal gradients and the rate of plate production. *Earth planet. sci. letters* **40**, 301–15.

Birch, F. 1961. Composition of the Earth's mantle. *Geophys. J. R. Astr. Soc.* **4**, 295–311.

Birch, F. 1968. On the possibility of large changes in the Earth's volume. *Phys. Earth planet. interiors* **1**, 141–7.

Birch, F., R. F. Roy and E. R. Decker 1968. Heat flow and thermal history in New England and New York. In *Studies of Appalachian geology*, E. Zen, W. S. White, J. B. Hadley and J. B. Thompson (eds), 437–51. New York: Wiley.

Black, D. C. 1978. Isotope anomalies in solar system material – what can they tell us? In *The origin of the Solar System*, S. F. Dermott (ed.), 583–96. Chichester: Wiley.

Black, R., R. Caby, A. Moussine-Pouchkine, R. Bayer, J. M. Bertrand, A. M. Boullier, J. Fabre and A. Lesquer 1979. Evidence for late Precambrian plate tectonics in West Africa. *Nature* **278**, 223–7.

Bolt, B. A. 1962. Gutenberg's early PKP observations. *Nature* **196**, 122–4.

Bolt, B. A. 1976. *Nuclear explosions and earthquakes.* San Francisco: W. H. Freeman.

Bott, M. H. P. 1971. *The interior of the Earth.* London: Edward Arnold.

Brandt, J. C. and S. P. Maran 1972. *New horizons in astronomy.* San Francisco: W. H. Freeman.

Briden, J. C. 1976. Application of palaeomagnetism to Proterozoic tectonics. *Phil. trans. R. Soc. Lond.* **280A**, 405–16.

Brooks, C., D. E. James and S. R. Hart 1976. Ancient lithosphere: its role in young continental lithosphere. *Science* **193**, 1086–94.

Brown, G. C. 1979. The changing pattern of batholith emplacement during Earth history. In *Origin of granite batholiths: geochemical evidence*, M. P. Atherton and J. Tarney (eds), 106–15. Orpington, Kent: Shiva.

Brown, G. C. and W. S. Fyfe 1972. The transition from metamorphism to melting: status of the granulite and eclogite facies. *Proc. 24th IGC, Montreal*, sect. 2, 27–34.

Brown, G. C. and J. Hennessy 1978. The initiation and thermal diversity of granite magmatism. *Phil. trans. R. Soc. Lond.* **288A**, 631–43.

219

Brown, G. C., J. A. Plant and R. S. Thorpe 1980. Plutonism in the British Caledonides: space, time and geochemistry, In *The Caledonides in the USA*, D. R. Wones (ed.), 157–66. Dept. of Geological Sciences, Virginia Polytechnic Institute and State University, Blacksburg, Virginia, Memoir No. 2.

Brown, G. M. 1978. Chemical evidence for the origin, melting, and differentiation of the Moon. In *The origin of the Solar System*, S. F. Dermott (ed.), 597–609. Chichester: Wiley.

Bukowinski, M. S. T. 1976. The effect of pressure on the physics and chemistry of potassium. *Geophys. res. letters*, **3**, 491–4.

Bullard, E. C. 1957. The density within the Earth. *Verh. geol. mijnb. genoot. ned. kolon* **18**, 23–41.

Bullard, E. C. 1971. The Earth's magnetic field and its origin. In *Understanding the Earth*, I. G. Gass, P. J. Smith and R. C. L. Wilson (eds), 71–80. Horsham, Sussex: Artemis Press.

Bullen, K. E. 1963. *Introduction to the theory of seismology*. Cambridge: Cambridge University Press.

Bullen, K. E. and R. A. W. Haddon 1970. Evidence from seismology and related sources on the Earth's present initial structure. *Phys. Earth planet. interiors* **2**, 342–9.

Burchfield, J. D. 1975. *Lord Kelvin and the age of the Earth*. London: Macmillan.

Burke, K. C. and J. T. Wilson 1976. Hot spots on the Earth's surface. *Scientific American* **235** (2), 46–57.

Busse, F. H. 1975. A model of the geodynamo. *Geophys. J.* **42**, 437–59.

Cameron, A. G. W. 1973. Abundances of the elements in the Solar System. *Space sci. rev.* **15**, 121–46.

Cann, J. R. 1970. New model for the structure of ocean crust. *Nature* **226**, 928–30.

Cann, J. R. 1974. A model for oceanic crustal structure developed. *Geophys. J. R. Astr. Soc.* **39**, 169–87.

Carmichael, I. S. E., F. J. Turner and J. Verhoogen 1974. *Igneous petrology*. New York: McGraw-Hill.

Carrigan, C. R. and D. Gubbins 1979. The source of the Earth's magnetic field. *Scientific American* **240** (2), 92–101.

Cathles, L. M. 1975. *The viscosity of the Earth's mantle*. Princeton, NJ: Princeton University Press.

Chapple, W. M. and T. E. Tullis 1977. Evaluation of the forces that drive plates. *J. geophys. res.* **82**, 1967–84.

Christensen, N. I. and M. H. Salisbury 1975. Structure and constitution of the lower oceanic crust. *Rev. geophys. space phys.* **13**, 57–86.

Chung, D. H. 1972. Birch's law: why is it so good? *Science* **177**, 261–3.

Clarke, S. P. 1971. *Structure of the Earth*. Englewood Cliffs, NJ: Prentice-Hall.

Clarke, S. P. and A. E. Ringwood 1964. Density distribution and constitution of the mantle. *Rev. geophys.* **2**, 35–88.

Clarke, S. P., K. Turekian and L. Grossman 1972. Model for early history of the Earth. In *The nature of the solid Earth*, E. C. Robertson (ed.), 3–18. New York: McGraw-Hill.

Clifford, T. N. 1970. The structural framework of Africa. In *African magmatism and tectonics*, T. N. Clifford and I. G. Gass (eds), 1–26. London: Oliver & Boyd.

Cole, G. H. A. 1978. *The structure of planets*. London: Wykeham.

Coleman, R. G. 1977. *Ophiolites: ancient oceanic lithosphere?* Heidelberg: Springer-Verlag.

Condie, K. C. 1976. *Plate tectonics and crustal evolution*. New York: Pergamon.

Cowie, J. W. 1974. The Cambrian of Spitsbergen and Scotland. In *Lower Palaeozoic rocks of the world*. Vol. 2: *Cambrian of the British Isles, Norden and Spitsbergen*, C. H. Holland (ed.), 123–56. Chichester: Wiley.

Davies, G. F. 1977a. Viscous mantle flow under moving lithospheric plates and under subduction zones. *Geophys. J. R. Astr. Soc.* **49**, 557–63.

Davies, G. F. 1977b. The roles of boundary friction, basal shear stress and deep mantle convection in plate tectonics. *Geophys. res. letters* **5**, 161–4.

Davies, G. F. 1977c. Whole mantle convection and plate tectonics. *Geophys. J. R. Astr. Soc.* **49**, 459–86.

Davies, G. F. 1979. Thickness and thermal history of continental crust and root zones. *Earth planet. sci. letters* **44**, 231–8.

Dermott, S. F. (ed.) 1978. *The origin of the Solar System.* Chichester: Wiley.

Dickinson, W. R. 1968. Circum-Pacific andesite types. *J. geophys. res.* **73**, 2261–9.

Drury, M. J. 1978. Partial melt in the asthenosphere: evidence from electrical conductivity data. *Phys. Earth planet. interiors* **17**, 16–20.

Duffield, W. A. 1972. A naturally occurring model of global plate tectonics. *J. geophys. res.* **77**, 2543–55.

Dunn, J. R., M. Fuller, M. Ito and V. A. Schmidt 1971. Palaeomagnetic study of a reversal of the Earth's magnetic field. *Science* **172**, 840–5.

Eaton, J. P. and K. T. Murata 1960. How volcanoes grow. *Science* **132**, 925–38.

Elsasser, W. M. 1963. Early history of the Earth. In *Earth science and meteoritics*, J. Geiss and E. D. Goldberg (eds), 1–30. Amsterdam: North-Holland.

Elsasser, W. M., P. Olson and B. D. Marsh 1979. The depth of mantle convection. *J. geophys. res.* **84**, 147–55.

Emslie, R. F. 1978. Anorthosite massifs, rapakivi granites, and late Proterozoic rifting of North America. *Precambrian res.* **7**, 61–98.

Ernst, W. G. 1969. *Earth materials.* Englewood Cliffs, NJ: Prentice-Hall.

Faber, S. and D. Bamford 1979. Lithospheric structural contrasts across the Caledonides of northern Britain. *Tectonophysics* **56**, 17–30.

Fisher, D. E. 1976. Rare gas clues to the origin of the terrestrial atmosphere. In *The early history of the Earth*, B. F. Windley (ed.), 547–56. Chichester: Wiley.

Forsyth, D. and S. Uyeda 1975. On the relative importance of the driving forces of plate motion. *Geophys. J. R. Astr. Soc.* **43**, 163–200.

Fryer, B. J., W. S. Fyfe and R. Kerrich 1979. Archaean volcanogenic oceans. *Chem. geol.* **24**, 25–33.

Fyfe, W. S. 1974. Archaean tectonics. *Nature* **249**, 338.

Fyfe, W. S. 1978. The evolution of the Earth's crust: modern plate tectonics to ancient hot spot tectonics? *Chem. geol.* **23**, 89–114.

Fyfe, W. S. 1979. The geochemical cycle of uranium. *Phil. trans. R. Soc. Lond.* **291A**, 433–45.

Fyfe, W. S. 1981. Andesites – the product of geosphere mixing. In *Andesites and related rocks*, R. S. Thorpe (ed.). Chichester: Wiley (in press).

Fyfe, W. S. and G. C. Brown 1972. Granite past and present. *J. Earth sciences* **8**, 249–60.

Garland, G. M. 1971. *Introduction to geophysics: mantle, crust and core.* Philadelphia: W. B. Saunders.

Gass, I. G. 1977. The evolution of the pan-African crystalline basement in NE Africa and Arabia. *J. Geol. Soc. Lond.* **134**, 129–38.

Gass, I. G. 1979. Magmagenesis. *Science progress* **65**, 251–68.

Gass, I. G. and J. D. Smewing 1980. Ophiolites: obducted oceanic lithosphere. In *The sea*, C. Emillani (ed.), Vol. 7, *The ocean lithosphere.* New York: Wiley (in press).

Gass, I. G., P. J. Smith and R. C. L. Wilson 1971. *Understanding the Earth.* Horsham, Sussex: Artemis.

Gillespie, C. C. 1951. *Genesis and geology.* Cambridge, Mass. Harvard University Press.

Goldreich, P. and W. R. Ward 1973. The formation of planetisimals. *Astrophys. J.* **183**, 1051–61.

Goldstein, J. I. and J. M. Short 1967. The iron meteorites: their thermal history and parent bodies. *Geochim. cosmochim. acta* **31**, 1733–70.

Gough, D. I. 1977. The geoid and single-cell mantle convection. *Earth planet. sci. letters* **34**, 360–4.

Green, D. A. 1975. On the postulated Hawaiian plume with emphasis on the limitations of seismic arrays for detecting deep mantle structure. *J. geophys. res.* **80**, 4028–36.

Green, D. H. and A. E. Ringwood 1963. Mineral assemblages in a model mantle composition. *J. geophys. res.* **68**, 937–45.

Greenwood, W. R., D. G. Hadley, R. E. Anderson, R. J. Fleck and D. L. Schmidt 1976. Late Proterozoic cratonisation in southwestern Saudi Arabia. *Phil. trans. R. Soc. Lond.* **280A**, 517–27.

Grossman, L. and J. W. Larimer 1974. Early chemical history of the Solar System. *Rev. geophys. space phys.* **12**, 71–101.

Gubbins, D. 1977. Energetics of the Earth's core. *J. geophys.* **43**, 453–64.

Gutenberg, B. 1959. *Physics of the Earth's interior.* New York: Academic Press.

Haber, F. C. 1959. *The age of the world – Moses to Darwin.* Baltimore: Johns Hopkins.

Hall, H. T. and T. Rama Murthy 1971. The early chemical history of the Earth: some critical elemental fractionations. *Earth planet. sci. letters* **11**, 239–44.

Hamilton, W. and W. B. Myers 1967. *The nature of batholiths.* US geol. surv. prof. paper, no. 554C, 30 pp.

Hargraves, R. B. 1976. Precambrian geologic history. *Science* **193**, 363–71.

Harris, P. G. 1971. The composition of the Earth. In *Understanding the Earth*, I. G. Gass, P. J. Smith and R. C. L. Wilson (eds), 53–70. Horsham, Sussex: Artemis.

Harris, P. G. and E. A. Middlemost 1970. The evolution of kimberlites. *Lithos* **3**, 79–90.

Hart, R. S., D. L. Anderson and H. Kanamori 1977. The effect of attenuation on gross Earth models. *J. geophys. res.* **82**, 1647–54.

Hawkesworth, C. J., M. J. Norry, J. C. Roddick, P. E. Baker, P. W. Francis and R. S. Thorpe 1979. $^{143}Nd/^{144}Nd$, $^{87}Sr/^{86}Sr$, and incompatible element variations in calc-alkaline andesites and plateau lavas from South America. *Earth planet. sci. letters* **42**, 45–57.

Heier, K. S. 1978. The distribution and redistribution of heat-producing elements in the continents. *Phil. trans. R. Soc. Lond.* **288A**, 393–400.

Heirtzler, J. R., G. O. Dickson, E. M. Herron, W. C. Pitman and X. Le Pichon 1968. Marine magnetic anomalies, geomagnetic field reversals, and motions of the ocean floor and continents. *J. geophys. res.* **73**, 2119–36.

Higgins, G. and G. C. Kennedy 1971. The adiabatic gradient and the melting point gradient in the core of the Earth. *J. geophys. res.* **76**, 1870–8.

Hoffman, P. 1973. Evolution of an early Proterozoic continental margin: the Coronation Geosyncline and associated aulacogens of the northwestern Canadian Shield. *Phil. trans. R. Soc. Lond.* **273A**, 547–81.

Huang, S. S. 1973. Extrasolar planetary systems. *Icarus* **18**, 339–76.

Ito, K. and G. C. Kennedy 1970. The fine structure of the basalt–eclogite transition. *Min. Soc. Am. spec. paper* **3**, 77–84.

Jacobs, J. A. 1953. The Earth's inner core. *Nature* **172**, 297–8.

Jacobs, J. A. 1975. *The Earth's core.* London: Academic Press.

Jacobs, J. A., R. D. Russell and J. T. Wilson 1974. *Physics and geology*, 2nd edn. New York: McGraw-Hill.

Jeffreys, H. 1962. *The Earth*, 4th edn. Cambridge: Cambridge University Press.

Johnston, D. H. and M. N. Toksöz 1977. Internal structure and properties of Mars. *Icarus* **32**, 73–84.

Jordan, T. H. 1975a. Lateral heterogeneity and mantle dynamics. *Nature* **257**, 745–50.

Jordan, T. H. 1975b. The continental tectosphere. *Rev. geophys. space phys.* **13**, 1–12.

Jordan, T. H. 1978. Composition and development of the continental tectosphere. *Nature* **274**, 544–8.

Jordan, T. H. 1979. The deep structure of the continents. *Scientific American* **240** (1), 70–82.

Julian, B. R., D. Davies and R. M. Sheppard 1972. PKJKP. *Nature* **235**, 317–18.

Kaula, W. M. 1972. Global gravity and mantle convection. *Tectonophysics* **13**, 341–59.

Kaula, W. M. 1975. The seven ages of a planet. *Icarus* **26**, 1–15.

Kaula, W. M. 1977. On the origin of the Moon, with emphasis on bulk composition. *Proc. 8th lunar sci. conf.* 321–31.

Kerridge, J. F. 1977. Iron: whence it came, where it went. *Space sci. rev.* **20**, 3–68.

Kerridge, J. F. 1978. Aspects of accretion in the early Solar System. In *The origin of the Solar System*, S. F. Dermott (ed.), 493–510. Chichester: Wiley.

Kittel, C. 1976. *Introduction to solid state physics*, 5th edn. New York: Wiley.

Knoll, A. H. and E. S. Barghoorn 1977. Archaean microfossils showing cell division from the Swaziland System of South Africa. *Science* **198**, 396–8.

Knopoff, L. 1972. Observation and inversion of surface wave data. *Tectonophysics* **13**, 497–519.

Kröner, A. 1979. Precambrian crustal evolution in the light of plate tectonics and the undation theory. *Geol. en mijnbouw* **58**, 231–40.

Kushiro, I. and H. S. Yoder 1966. Anorthite-forsterite and anorthite–enstatite reactions and their bearing on the basalt–eclogite transformation. *J. petrol.* **7**, 337–62.

Lachenbruch, A. H. 1970. Crustal temperatures and heat production: implications of the linear heat flow relation. *J. geophys. res.* **75**, 3291–300.

Lachenbruch, A. H. 1973. A simple mechanical model for oceanic spreading centres. *J. geophys. res.* **78**, 3395–417.

Laporte, F. 1979. *Ancient environments*, 2nd edn. Englewood Cliffs, NJ: Prentice-Hall.

Lee, T., D. A. Papanastassiou and G. J. Wasserburg 1978. Calcium isotopic anomalies in the Allende meteorite. *Astr. J.* **220**, L21–L25.

Lee, W. H. K. and S. Uyeda 1965. Review of heat flow data. In *Terrestrial heat flow*, W. H. K. Lee (ed.): Am. Geophys. Union geophysical monograph, no. 8.

Leggett, J. K., W. S. McKerrow, J. H. Morris, G. J. H. Oliver and W. E. A. Phillips 1979. The north-western margin of Iapetus. In *The Caledonides of the British Isles – reviewed*, A. L. Harris, C. H. Holland and B. E. Leake (eds.) Geol. Soc. Lond. spec. pub. no. 8, 499–512.

Le Pichon, X., J. Franchetau and J. Bonnin 1973. *Plate tectonics*. Amsterdam: Elsevier.

Lewis, J. S. 1971. Consequences of the presence of sulphur in the core of the Earth. *Earth planet. sci. letters* **11**, 130–4.

Lewis, J. S. 1972. Metal–silicate fractionation in the Solar System. *Earth planet. sci. letters* **15**, 286–90.

Liebermann, R. C. and A. E. Ringwood 1973. Birch's law and polymorphic phase transformations. *J. geophys. res.* **78**, 6926–32.

Loper, D. E. 1975. Torque balance and energy budget for the precessionally driven dynamo. *Phys. Earth planet. interiors* **11**, 43–60.

Loper, D. E. 1978. Some thermal consequences of a gravitationally powered dynamo. *J. geophys. res.* **83**, 5961–70.

Lowes, F. J. 1970. Possible evidence on core evolution from geomagnetic dynamo theories. *Phys. Earth planet. interiors* **2**, 383–5.

McAlester, A. L. 1977. *The history of life*, 2nd edn. Englewood Cliffs, NJ: Prentice-Hall.

McCall, G. J. H. 1973. *Meteorites and their origins*. Newton Abbot: David & Charles.

McCrea, W. H. 1963. The origin of the Solar System. *Contemp. phys.* **4**, 278–90.

McCrea, W. H. 1974. Origin of the Solar System: review of concepts and theories. In *On the origin of the Solar System*, H. Reeves (ed.), 2–20. Conf. rep., Nice, 3–7 April 1972. Paris: CNRS.

McCrea, W. H. 1978. The formation of the Solar System: a protoplanet theory. In *The origin of the Solar System*, S. F. Dermott (ed.), 75–110. Chichester: Wiley.

McElhinny, M. W. 1973. *Palaeomagnetism and plate tectonics*. Cambridge: Cambridge University Press.

McElhinny, M. W. and B. J. J. Embleton 1976. Precambrian and early Palaeozoic palaeomagnetism in Australia. *Phil. trans. R. Soc. Lond.* **280A**, 417–32.

McGregor, A. M. 1951. Some milestones in the Precambrian of southern Africa. *Proc. Geol. Soc. S. Afr.* **54**, 27–71.

McKenzie, D. P. and F. Richter 1976. Convection currents in the Earth's mantle. *Scientific American* **235** (5), 72–89.

McKenzie, D. P. and N. Weiss 1975. Speculations on the thermal and tectonic history of the Earth. *Geophys. J. R. Astr. Soc.* **42**, 131–74.

McNutt, R. H., J. H. Crocket, A. H. Clark, J. C. Caelles, E. Farrar, S. J. Haynes and M. Zentilli 1975. Initial $^{87}Sr/^{86}Sr$ ratios of plutonic and volcanic rocks of the central Andes between latitudes 26° and 29° south. *Earth planet. sci. letters* **27**, 305–13.

Mason, B. 1966. *Principles of geochemistry*, 3rd edn. New York: Wiley.

Mather, K. F. and S. L. Mason 1939. *A source book in geology.* New York: McGraw-Hill.

Matthews, D. H., J. Lort, T. Vertue, C. K. Poster and I. G. Gass 1971. Seismic velocities at the Cyprus outcrop. *Nature phys. sci.* **231**, 200–1.

Metchnik, V. I., M. T. Gladwin and F. D. Stacey 1974. Core convection as a power source for the geomagnetic dynamo: a thermodynamic argument. *J. geomagn. geoelec.* **26**, 405–15.

Minster, J. B. and T. H. Jordan 1978. Present-day plate motions. *J. geophys. res.* **83**, 5331–54.

Minster, J. B., T. H. Jordan, P. Molnar and E. Haines 1974. Numerical modelling of instantaneous plate tectonics. *Geophys. J. R. Astr. Soc.* **36**, 541–76.

Miyashiro, A. 1972. Metamorphism and related magmatism in plate tectonics. *Amer. J. sci.* **272**, 629–56.

Molnar, P. and D. Gray 1979. Subduction of continental lithosphere: some constraints and uncertainties. *Geology* **7**, 58–62.

Moorbath, S. 1977. Ages, isotopes and evolution of the Precambrian continental crust. *Chem. geol.* **20**, 151–87.

Moorbath, S. 1978. Age and isotope evidence for the evolution of the continental crust. *Phil. trans. R. Soc. Lond.* **288A**, 401–12.

Morgan, W. J. 1971. Convection plumes in the lower mantle. *Nature* **230**, 42–3.

Morgan, W. J. 1972. Deep mantle convection and plate motions. *Bull. Am. Ass. Petrol. Geol.* **56**, 203–13.

Myers, J. S. 1975. Vertical crustal movements of the Andes in Peru. *Nature* **254**, 672–4.

Nagy, B. 1975. *Carbonaceous meteorites.* Amsterdam: Elsevier.

O'Nions, R. K., N. M. Evensen, P. J. Hamilton and S. R. Carter 1978. Melting of the mantle past and present: isotope and trace element evidence. *Phil. trans. R. Soc. Lond.* **288A**, 547–59.

O'Nions, R. K., N. M. Evensen and P. J. Hamilton 1980. Differentiation and evolution of the mantle. *Phil. trans. R. Soc. Lond.* (in press).

Oxburgh, E. R. and D. L. Turcotte 1974. Thermal gradients and regional metamorphism in overthrust terrains with special reference to the eastern Alps. *Schweiz. Mineral. Petrog.* **54**, 641–62.

Pauling, L. 1959. *The nature of the chemical bond*, 3rd edn. Oxford: Oxford University Press.

Phillips, W. E. A., C. J. Stillman and T. Murphy 1976. A Caledonian plate tectonic model. *J. Geol. Soc. Lond.* **132**, 579–609.

Piper, J. D. A. 1976. Palaeomagnetic evidence for a Proterozoic supercontinent. *Phil. trans. R. Soc. Lond.* **280A**, 469–90.

Piper, J. D. A., J. C. Briden and K. Lomax 1973. Precambrian Africa and S. America as a single continent. *Nature* **245**, 244–8.

Pitcher, W. S. 1978. The anatomy of a batholith. *J. Geol. Soc. Lond.* **135**, 157–82.

Pollack, H. N. and D. S. Chapman 1977. Mantle heat flow. *Earth planet sci. letters* **34**, 174–84.

Prentice, A. J. R. 1978. Towards a modern Laplacian theory for the formation of the Solar System. In *The origin of the Solar System*, S. F. Dermott (ed.), 111–61. Chichester: Wiley.

Press, F. 1968. Earth models obtained by Monte Carlo inversion. *J. geophys. res.* **73**, 5223–34.

Press, F. 1970a. Earth models consistent with geophysical data. *Phys. Earth. planet. interiors* **3**, 3–22.

Press, F. 1970b. Regionalized Earth models. *J. geophys. res.* **75**, 6575–8.

Press, F. and R. Siever 1978. *Earth*, 2nd edn. San Francisco: W. H. Freeman.

Qamar, A. 1973. Revised velocities in the Earth's core. *Bull. Seism. Soc. Am.* **63**, 1073–105.

Reeves, H. (ed.) 1974. *On the origin of the Solar System*. Conf. rep., Nice, 3–7 April 1972. Paris: CNRS.

Reeves, H. 1978. The origin of the Solar System. In *The origin of the Solar System*, S. F. Dermott (ed.), 1–17. Chichester: Wiley.

Richter, C. F. 1958, *Elementary seismology*. San Francisco: W. H. Freeman.

Ringwood, A. E. 1969. Phase transformations in the mantle. *Earth planet. sci. letters* **5**, 401–12.

Ringwood, A. E. 1974. The petrological evolution of island arc systems. *J. Geol. Soc. Lond.* **130**, 183–204.

Ringwood, A. E. 1975. *Composition and petrology of the Earth's mantle*. New York: McGraw-Hill.

Ringwood, A. E. 1977a. *Composition and origin of the Earth*. Res. School of Earth Sciences, ANU pub., No. 1299.

Ringwood, A. E. 1977b. Composition of the core and implications for the origin of the Earth. *Geochem. J.* **11**, 111–35.

Ringwood, A. E. and D. L. Anderson 1977. Earth and Venus: a comparative study. *Icarus* **30**, 243–53.

Ringwood, A. E. and S. P. Clark 1971. Internal constitution of Mars. *Nature* **234**, 89–92.

Ringwood, A. E. and A. Major 1970. The system Mg_2SiO_4–Fe_2SiO_4 at high pressures and temperatures. *Phys. Earth planet. interiors* **3**, 89–108.

Ronov, A. B. and A. A. Yaroshevsky 1969. Chemical composition of the Earth's crust. In *The Earth's crust and upper mantle*, P. J. Hart (ed.), 37–57. Am. Geophys. Union.

Runcorn, S. K. (ed.) 1967. *Dictionary of geophysics*. Oxford: Pergamon.

Schopf, J. W. 1978. The evolution of the earliest cells. *Scientific American* **239** (3), 84–102.

Schramm, D. N. and R. N. Clayton 1978. Did a supernova trigger the formation of the solar system? *Scientific American* **239** (4), 98–113.

Schubert, G., D. A. Yuen and D. L. Turcotte 1975. Role of phase transitions in a dynamic mantle. *Geophys. J. R. Astr. Soc.* **42**, 705–35.

Sclater, J. G. and J. Crowe 1979. A heat flow survey at anomaly 13 on the Reykjanes Ridge: a critical test of the relation between heat flow and age. *J. geophys. res.* **84**, 1593–602.

Sclater, J. G. and J. Franchetau 1970. The implications of terrestrial heat flow observations on current tectonic and geochemical models of the crust and upper mantle of the Earth. *Geophys. J. R. Astr. Soc.* **20**, 509–42.

Sclater, J. G., C. Jaupart and D. Galson 1980. The heat flow through oceanic and continental crust and the heat loss of the Earth. *Rev. geophys. and space phys.* **18**, 269–311.

Shimizu, M. 1976. Instability of a highly reducing atmosphere on the primitive Earth. *Precambrian res.* **3**, 463–70.

Shklovskii, I. S. 1978. *Stars: their birth, life and death*. San Francisco: W. H. Freeman.

Simpson, P. R., G. C. Brown, J. A. Plant and D. Ostle 1979. Uranium mineralisation and granite magmatism in the British Isles. *Phil. trans. R. Soc. Lond.* **291A**, 385–412.

Smith, J. V. 1979. Mineralogy of the planets: a voyage in space and time. *Min. mag.* **43**, 1–89.

Smith, P. J. 1973. *Topics in geophysics*. Milton Keynes: Open University Press.

Spitzer, L. 1939. The dissipation of planetary filaments. *Astrophys. J.* **90**, 675–88.

Stacey, F. D. 1977. *Physics of the Earth*, 2nd edn. New York: Wiley.

Stocker, R. L. and M. F. Ashby 1973. On the rheology of the upper mantle. *Rev. geophys. space phys.* **11**, 391–426.

Talwani, M., X. Le Pichon and M. Ewing 1965. Crustal structure of the mid-ocean ridges. 2: Computed model from gravity and seismic refraction data. *J. geophys. res.* **70**, 341–52.

Tarling, D. H. 1978. *Evolution of the Earth's crust*. London: Academic Press.

Tarney, J., I. W. D. Dalziel and M. J. de Wit 1976. Marginal basin 'Rocas Verdes' complex from S. Chile: a model for Archaean greenstone belt formation. In *The early history of the Earth*, B. F. Windley (ed.), 131–46. Chichester: Wiley.

Tayler, R. J. 1972a. *The origin of the chemical elements.* London: Wykeham.

Tayler, R. J. 1972b. *The stars: their structure and evolution.* London: Wykeham.

Taylor, S. R. 1963. Trace element abundances and the chondritic Earth model. *Geochim. cosmochim. acta* **28**, 1989–98.

Taylor, S. R. 1975. *Lunar science: a post-Apollo view.* New York: Pergamon.

Ter Haar, D. 1948. Studies on the origin of the Solar System. *Proc. R. Danish Acad. Sci.* **25**, no. 3.

Ter Haar, D. 1950. Further studies on the origin of the Solar System. *Astr. J.* **111**, 179–90.

Thorpe, R. S. (ed.) 1981. *Orogenic andesites and related rocks.* Chichester: Wiley.

Thorpe, R. S. and P. W. Francis 1979. Variations in Andean andesite compositions and their petrogenetic significance. *Tectonophysics* **57**, 53–70.

Toksöz, M. N. 1976. The subduction of lithosphere. In *Continents adrift and aground*, J. T. Wilson (ed.), 113–22. San Francisco: W. H. Freeman.

Towe, K. M. 1978. Early Precambrian oxygen: a case against photosynthesis. *Nature* **274**, 657–61.

Trimble, V. 1975. The origin and abundances of the chemical elements. *Rev. mod. phys.* **47**, 877–976.

Turcotte, D. L. and E. R. Oxburgh 1973. Mid-plate tectonics. *Nature* **244**, 337–9.

Turcotte, D. L. and E. R. Oxburgh 1978. Intra-plate volcanism. *Phil. trans. R. Soc. Lond.* **288A**, 561–79.

Turekian, K. K. 1972. *Chemistry of the Earth.* New York: Holt, Rinehart & Winston.

Turner, F. J. 1968. *Metamorphic petrology.* New York: McGraw-Hill.

Turner, J. S. 1973. Convection in the mantle: a laboratory model with temperature dependent viscosity. *Earth planet. sci. letters* **17**, 369–74.

Tuttle, O. F. and N. L. Bowen 1958. Origin of granite in the light of experimental studies in the system $NaAlSi_3O_8$–$KAlSi_3O_8$–SiO_2–H_2O. *Geol. Soc. Am. Mem.* **74**.

Urey, H. C. 1952. *The planets: their origin and development.* New Haven, Conn.: Yale University Press.

Usselman, T. N. 1975. Experimental approach to the state of the core. I: The liquidus relations of the Fe-rich portion of the Fe–Ni–S system from 30 to 100 kb. *Am. J. sci.* **275**, 291–303.

Uyeda, S. 1978. *The new view of the Earth.* San Francisco: W. H. Freeman.

Van Schmus, W. R. and J. A. Wood 1967. A chemical petrologic classification for the chondritic meteorites. *Geochim. cosmochim. acta* **31**, 747–65.

Veizer, J. 1976. $^{87}Sr/^{86}Sr$ evolution of seawater during geologic history and its significance as an index of crustal evolution. In *The early history of the Earth*, B. F. Windley (ed.), 569–78. London: Wiley.

Verhoogen, J. 1961. Heat balance of the Earth's core. *Geophys. J. R. Astr. Soc.* **4**, 276–81.

Verhoogen, J. 1973. Thermal regime of the Earth's core. *Phys. Earth planet. interiors* **7**, 47–58.

Vidal, P. and L. Dosso 1978. Core formation: catastrophic or continuous? Sr and Pb isotope geochemistry constraints. *Geophys. res. letters* **5**, 169–72.

Vollmer, R. 1977. Terrestrial lead isotopic evolution and formation time of the Earth's core. *Nature* **270**, 144–7.

Wang, C. Y. 1972. A simple Earth model. *J. geophys. res.* **77**, 4318–29.

Wasson, J. T. 1974. *Meteorites: classification and properties.* Heidelberg: Springer-Verlag.

Watson, J. V. 1978. Precambrian thermal regimes. *Phil. trans. R. Soc. Lond.* **288A**, 431–40.

Wiik, H. B. 1956. The chemical composition of some stony meteorites. *Geochim. cosmochim. acta* **9**, 279–89.

Williams, D. L. and R. P. Von Herzen 1974. Heat loss from the Earth: new estimate. *Geology* **2**, 327–8.

Williams, I. P. 1975. *The origin of the planets*. Bristol: Adam Hilger.

Williams, I. P. and A. W. Cremin 1968. A survey of theories of the origin of the Solar System. *Quart. J. R. Astr. Soc.* **9**, 40–62.

Williamson, E. D. and L. H. Adams 1923. Density distribution in the Earth. *J. Wash. Acad. Sci.* **13**, 413–28.

Wilson, J. F., M. J. Bickle, C. J. Hawkesworth, A. Martin, E. G. Nisbet and J. L. Orpen 1978. Granite–greenstone terrains of the Rhodesian Archaean craton. *Nature* **271**, 23–7.

Wilson, J. T. 1963. A possible origin of the Hawaiian islands. *Can. J. phys.* **41**, 863–70.

Windley, B. F. 1977. *The evolving continents*. Chichester: Wiley.

Windley, B. F. 1980. Evidence for land emergence in the early-to-middle Precambrian. *Proc. Geol. Ass.* **91**, 13–23.

Wood, J. A. 1968. *Meteorites and the origin of the planets*. New York: McGraw-Hill.

Wood, J. A. 1978. Origin of the Earth's moon. In *Planetary satellites*, J. A. Burns (ed.), 513–33. Tucson: University of Arizona Press.

Wood, J. A. 1979. *The Solar System*. Englewood Cliffs, NJ: Prentice-Hall.

Woolfson, N. M. 1978a. The capture theory and the origin of the Solar System. In *The origin of the Solar System*, S. F. Dermott (ed.), 179–98. Chichester: Wiley.

Woolfson, N. M. 1978b. The evolution of the Solar System. In *The origin of the Solar System*, S. F. Dermott (ed.), 199–217. Chichester: Wiley.

Worthington, M. H., J. R. Cleary and R. S. Anderssen 1972. Density modelling by Monte Carlo inversion. II: Comparison of recent Earth models. *Geophys. J. R. Astr. Soc.* **29**, 445–57.

Wyllie, P. J. 1971. *The dynamic Earth*. New York: Wiley.

Wyllie, P. J. 1976. *The way the Earth works*. New York: Wiley.

227

Index

Pages shown in bold figures contain the main definition or explanation of the term where this is relevant.

233

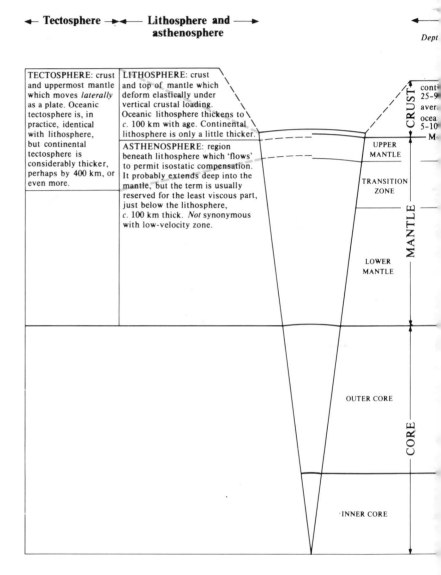

TECTOSPHERE: crust and uppermost mantle which moves *laterally* as a plate. Oceanic tectosphere is, in practice, identical with lithosphere, but continental tectosphere is considerably thicker, perhaps by 400 km, or even more.

LITHOSPHERE: crust and top of mantle which deform elastically under vertical crustal loading. Oceanic lithosphere thickens to *c.* 100 km with age. Continental lithosphere is only a little thicker.

ASTHENOSPHERE: region beneath lithosphere which 'flows' to permit isostatic compensation. It probably extends deep into the mantle, but the term is usually reserved for the least viscous part, just below the lithosphere, *c.* 100 km thick. *Not* synonymous with low-velocity zone.

CRUST

cont 25–9 aver ocea 5–10 M

UPPER MANTLE

TRANSITION ZONE

LOWER MANTLE

MANTLE

OUTER CORE

INNER CORE

CORE

Crust, mantle and core →

, km	Density, kg m^{-3} (× 10³)	% of Earth's mass	Seismic velocities V_p and V_s, km s^{-1}	Physical properties of regions	Compositions of regions	Nature of boundaries
	0 5 10 15		0 5 10 15			
ental:		0·7%		Defined seismically as the region above the Mohorovičić discontinuity (Moho).	CONTINENTAL: granodiorite overlying intermediate to basic granulite. OCEANIC: basalt.	Conrad discontinuity beneath continents due to granodiorite → granulite change.
ge: 35						
ic:						
ho				LOW VELOCITY ZONE: few % partial melt; more prominent below oceans.	DEPLETED PERIDOTITE ↓ GARNET LHERZOLITE	Compositional
— 400				SOLID (apart from low-velocity zone): elastic to passage of seismic waves, but solid-state creep permits inelastic deformation beneath lithosphere (q.v.) and convection beneath tectosphere (q.v.)	Development of high-pressure phases, but still with composition of garnet lherzolite.	PHASE CHANGE TO HIGH-PRESSURE POLYMORPHS
- 1050		68·3%	V_s V_p			
- 2885						Compositional, resulting in solid → liquid
		31%	V_p	LIQUID CONVECTING SOURCE OF MAGNETIC FIELD	IRON-SULPHUR MIXTURE (c. 86% Fe, 12% S, 2% Ni)	
5155 —						Compositional, resulting in liquid → solid
		1·7%	V_s V_p	SOLID	IRON-NICKEL ALLOY	
6370 —